P9-DGI-294

SUPER VOLCANOES

SUPER VOLCANOES

What They Reveal about Earth and the Worlds Beyond

ROBIN GEORGE ANDREWS

W. W. NORTON & COMPANY
Independent Publishers Since 1923

Printed in the United States of America
First Edition

For information about permission to reproduce selections from this book, write to
Permissions, W. W. Norton & Company, Inc., 500 Fifth Avenue, New York, NY 10110

For information about special discounts for bulk purchases, please contact
W. W. Norton Special Sales at specialsales@wwnorton.com or 800-233-4830

Manufacturing by Lake Book Manufacturing
Production manager: Beth Steidle

Library of Congress Cataloging-in-Publication Data

Names: Andrews, Robin George, author.
Title: Super volcanoes : what they reveal about earth and the worlds beyond /
Robin George Andrews.
Description: First edition. | New York, NY : W.W. Norton & Company, [2022] |
Includes bibliographical references and index.
Identifiers: LCCN 2021031293 | ISBN 9780393542066 (hardcover) |
ISBN 9780393542073 (epub)
Subjects: LCSH: Super volcanoes. | Volcanoes.
Classification: LCC QE522 .A56 2022 | DDC 551.21—dc23
LC record available at https://lccn.loc.gov/2021031293

W. W. Norton & Company, Inc., 500 Fifth Avenue, New York, N.Y. 10110
www.wwnorton.com

W. W. Norton & Company Ltd., 15 Carlisle Street, London W1D 3BS

1 2 3 4 5 6 7 8 9 0

For the Nans, Nanu, and Grandad Charlie

CONTENTS

PROLOGUE

THE GATE IN THE SKY

Until that day, I had never been sat on by a cloud. It came as quite a shock: I had briefly fallen asleep, or at least thought I had, on the jet-black pebble-covered ridge of the mountain. I had dissociated my exhausted self from my surroundings and had drifted off, lost in the sky. I was only brought shuddering back to reality when a collection of condensing water, which quite frankly should have been minding its own business several hundred feet downslope, was given a helping hand by a persistent blast of wind. It tumbled up the steep flank nearby, bubbled up over the ridge, and came to a stop on my head. It was like I had suddenly perspired ice; I shuddered as my neurons fizzed with surprise. Mischief managed, the cloud promptly burst and faded away along the thin rocky path.

"Joe," I said, calling out in the twilight. "Joe?"

Something stirred next to me. "Yeah?"

"A cloud just ambushed me. It sat on me. I think."

"Okay," he said, rising to his feet. "That does it." He winced. By the look of his crinkling face, his limbs, like mine, ached a lot more than he had anticipated. Joe emphatically jammed his walking pole into the crunchy ground. "Let's keep going," he spluttered. He wasn't about to become similarly soaked.

About nine hours earlier, we were wading through humid air and being poached in 104-degree Fahrenheit Tokyo heat. But now, the mercury had plunged to 14 degrees. Enveloped by an old US Navy coat and several extra layers underneath, I could still feel the chilly aeolian blades cutting me into ribbons. The air up here was also getting increasingly less packed with delicious oxygen, so breathing took a little more effort than usual. Joe's thermal shielding was more hastily assembled: he was adorned with a woolen, buttoned-up shirt and as many T-shirts as he could fit on in one go. We were fueled by little more than water and the 20 or so seaweed rice balls we hastily purchased earlier that day at a minimart in Akihabara, a hyperluminous Elysium of video games, manga, cat cafés, bleeps and bloops.

It was 2013. I was more than halfway through my PhD, something I was working on in New Zealand. Along with the usual storm of anxiety that most people experience while trying to put together a doctoral thesis, I was also thousands of miles from the vast majority of my friends and family in a country that, although an aesthetic wonder, turned out to be a little too tranquil for my liking—not the sort of place I was content living in for several years. I stuck it out, but loneliness was a frequent visitor.

Every now and then, I managed to escape and travel beyond the bounds of that far-flung archipelago. In 2013, a volcano-centric scientific conference was taking place in Japan, way out west in a bustling city named Kagoshima—the perfect excuse to spend some time exploring my favorite place on Earth. After the academic gathering, I spent the next few weeks wandering through forested islands taken right out of a Studio Ghibli movie, getting flown over viridian bowls and peaks in a helicopter, weaving through neon-illuminated drink-filled *izakaya*, ducking into crimson shrines to hide from the thumping, glorious rain, trying to break the top-100 score rankings on a rhythm-action game in the arcades, slurping ramen noodles enthusiastically, and going to a madcap music festival. To my enormous relief, I felt my spirits were sky-high. I was alone, but no longer lonely. My mind was erupting with newfound energy.

I also, perhaps unwisely, decided to climb Mount Fuji. This mountain is famed for its staggeringly beautiful symmetry. During a few weeks in the summer, when the ice is less pervasive, you can climb it. The climb is not particularly hazardous, but it gets steep, rocky, and extremely cold near the summit, 12,400 feet up. I thought I was making the ascent alone until, out of the blue, my childhood friend Joe—who had not long ago quit his job and had traveled to Thailand to learn kickboxing, because why not—had suddenly turned up in Tokyo after hearing I was there. His arrival proved to be a vaccine for melancholia.

After a few days of gallivanting about the capital, we both decided that we would clamber up Fuji together. The only problem, aside from Joe's lack of proper clothing and shoes, was that the only day we could head up the mountain was the day before Muse, our favorite band, was playing at the festival. It was going to be a close call, but being decidedly risk-prone, we steeled ourselves, caught a bus to one of the stations on the lower flanks of the mountain, and began.

After a few hours, the treetops gave way to the clouds, which passed below us as the Sun slipped over the edge of the planet. The light of sparkling cities failed to break through the thick blanket of weather downslope, permitting the night sky to explode above us.

We eventually reached a small collection of huts. The owners of one gifted us a delectable Japanese curry and, afterward, a set of pillows for a 90-minute nap, allowing our lungs to adjust as quickly as they could to the thinning atmosphere. We stepped outside and continued our climb, when a streak of light caught my eye. Then another. And another. And another!

Until then, it had completely passed me by that we were climbing Mount Fuji during the peak of the Perseid meteor shower. With no artificial light to sully the night, we had a front-row seat to the combustion of hundreds of fragments of a lonesome comet, hiding on the fringes of our cosmic neighborhood.

Soon after, an inky blue paint began to splash itself along the sky's floor. It hit us that the Sun was rising, fast. We were told by another group of climbers on the way up that the path thins out near the summit. There, climbers have to wait in a heavenly queue, taking turns to step through the shrine gate that marks the entrance to the mountain's peak. If we lagged behind, we would be stuck up there too long to make it all the way back down to the festival.

We began to run. The next few hours were a blur. I remember stepping past people who had begun to faint, tripping over the increasingly sharp rocks, watching as our previously invisible pathway up became flecked with reflected light. We paused for just a few moments when our legs felt like they were on fire, lying down on a ridge away from the path that, as it turns out, was the perfect spot for a cloud to ambush us. With five minutes to spare, and with acute breathlessness, we made it through the gate and collapsed on a ridge at the very top of the mountain. The drastically reduced air temperature and pressure made us both feel like zombies, but we didn't care. We made it—and for the only time in my life, I could feel what seemed like a single beam of sunlight on my face as the Sun crested the horizon. All the climbers who had summited the mountain applauded in collective respect and relief.

Against all odds, we also made it to the gig on the outskirts of Tokyo later that evening. We even found the energy for a few celebratory drinks beforehand. It remains one of the most exhilarating 24 hours of our lives. And we both owed much of it to that majestic mountain, that staircase to the Sun, cometary fires, and alien stars.

But it isn't just a mountain. It wasn't squashed together over eons by two slabs of rock crashing into each other in slow motion. It is the creation of epochs of eruptions, a monument to magma. Mount Fuji is a sleeping volcano, one that grew from the earth and rose to become a perfect otherworldly pedestal to the cosmos. That knowledge, always in the back of my mind as we relentlessly put one foot in front of the other, feels less like a fact of science to me and more like a consequence of magic.

INTRODUCTION

Volcanoes have a terrible reputation. When most adults think of volcanoes, especially adults that don't live particularly close to one, they think of huge explosions that often result in death and destruction. Volcanoes are perceived as harbingers of doom whose sole purpose is to maim and melt.

It isn't difficult to work out where this reputation comes from. Science journalists, like myself, write about volcanoes whenever they do something noteworthy or when scientists poke around them a bit and find out something fascinating or surprising about them. But volcanoes understandably crop up in the general news cycle when they erupt near a bunch of people and those people are put in peril. If an eruption ends up killing people, as happens every so often, we are treated to grim tales of suffering and some horrific accompanying imagery. These stories tend to be punctuated by two common threads. The first features people wondering aloud why no one saw this coming. The second involves scientists doing their utmost to explain that no volcano telegraphs whether it will erupt, precisely when it will erupt, and in what manner it will erupt. And these stories often give the impression that volcanoes are scary, unpredictable bombs hiding in cones of immense danger.

Volcano soothsayers on social media and unscrupulous editors at terrible news outlets, on the prowl for clicks, will make bold and entirely false claims in order to get a good harvest. They will tell you that an innocuous volcano is without question about to erupt and destroy this city/country/world. They will imply scientists have no idea what is happening or that they do know but are trying to hide the truth from the masses. As the past few years have made nauseatingly clear, misinformation playing to people's fears quickly becomes viral. Millions of people, anxious about the state of the world, get their information on volcanoes through the prognostications of these charlatans.

And then, of course, there's Hollywood. Don't get me wrong: if it's a good film, you can use volcanoes to spectacular effect. But the vast majority of movies that feature volcanoes frame them as nature's ovens of annihilation. Whether you are trying to melt Chris Pratt and some dinosaurs or are hoping to destroy an evil ring, volcanoes will do the job. It doesn't have to be lava that puts our characters in danger. Volcanoes have a bottomless box of deadly tricks, from avalanches of superheated matter moving at supersonic speeds to showers of asphyxiating ash. And because volcanoes have a phenomenal sense of dramatic timing, they will always do the most inconvenient, visually over-the-top thing possible whenever it is required of them. You can almost hear the volcano laughing like a stereotypical comic book supervillain as it keeps chasing screaming humans around. There are exceptions, but movies largely treat volcanoes like supremely powerful antagonists: they are there to end lives, burn down cities, and even bring an end to entire civilizations.

Okay. To be fair, volcanoes have done all three of these things.

A scientific paper published in 2017 had the unenviable task of estimating how many people have been killed by volcanoes, directly (for example, hit by lava) or indirectly (for example, severely altering the weather to cause a famine), in the past 500 years. Although counting the dead gets more difficult the further back in time you go, this reliable record came to a tally of 278,368 fatalities.[1] Some of these deaths took place on the slopes of a volcano, while others happened hundreds of miles away.

Eruptions can sometimes jettison all kinds of sunlight-mangling vapors into the sky that end up briefly but acutely disrupting weather patterns, causing some parts of the world to suddenly dry up while others get a soaking. On a few historical occasions this has triggered environmental troubles so unyielding and severe that teetering empires were finally pushed over the edge, including the Roman Republic[2] and the Ptolemaic dynasty in Ancient Egypt.[3]

On a planetary scale, they've done far worse. Around 252 million years ago, a planet already suffering from ecological turmoil was also baking thanks to the ~2-million-year-long eruption of lava gushing out of what is now Siberia. This continental-scale volcanism, unleashing climate-perturbing gases of its own, also ignited a huge reservoir of coal, triggering a global warming offensive. When all was said and done, this *Murder on the Orient Express*–style apocalypse killed more than nine out of every ten marine species and seven out of ten terrestrial vertebrate species—birds, amphibians, reptiles, mammals, and so forth—on the planet. This event, aptly known as the Great Dying, was easily the worst mass extinction in Earth's history and its darkest chapter.[4] Life barely made it through to the other side.

But here's the thing: volcanoes spend most of their time not erupting. When they do erupt, they often don't kill anyone. On average, there are about 40 volcanoes on Earth spewing lava or ash at any single moment in time. They don't usually make headlines because they are merely obeying the laws of thermodynamics, not melting people.

Volcanoes, like hurricanes, tornadoes, earthquakes, and so on, aren't inherently dangerous. They become a danger when people are in the way of them. We, the people, make the hazard. Eruptions only kill us because we, knowingly and unknowingly, built our cities on the slopes of lava factories.

Around 800 million people live within 60 miles or so of an active volcano.[5] The reasons why are complex. The land closer to the riskier sectors of a volcano is often cheaper and is therefore more likely to be lived on by those lower down on the socioeconomic ladder. There are also plenty of grim historical reasons why people live close to potentially lethal volcanoes. Jazmin Scarlett, a social and historical volcanologist, once shared with me a particularly egregious example. Before the Europeans arrived, the indigenous population of the West Indies' island of St. Vincent lived along the shores. But after becoming a British colony in the 1760s, the newly enslaved people were forced to live much closer to the island's La Soufrière volcano. A powerful eruption in 1812 killed many slaves working on inland plantations. Another major eruption in 1902–1903 left the liberated descendants of former slaves in financial ruin. Today, those living closest to the volcano are often the island's poorest.

But, for a decent proportion of those 800 million, living with a volcano on the horizon or right on their doorstep is a choice.[6] Although often aware of the hazards it poses, their lives entwine with the volcano's benefits. Volcanoes can provide fertile soil to grow crops on, a center of religious or spiritual importance, a nexus for tourism, a hot spot of biodiversity, and an astoundingly beautiful backdrop to people's homes. For many, the volcano is their home. The risk of a dangerous eruption, which varies from volcano to volcano, is usually an acceptable price to pay. People in Florida may wonder why anyone would choose to live near an active volcano, but the inhabitants of a volcanic region may wonder why anyone would choose to live in a part of the world that gets hit every year by hurricanes, which are only becoming more powerful because of climate change.

I'm not pretending that volcanoes can't be dangerous. Nearly 280,000 deaths is *a lot* of fatalities, and in some cases, tens of thousands of people can be slaughtered by a single eruption. But as I write this in the fading summer of 2020, a respiratory virus that barely registered six months ago has already killed nearly a million people all over the world. Unlike volcanoes, which have far more benefits than drawbacks, deadly respiratory viruses don't have any plus sides. They are thoroughly terrible.

I WOULD ARGUE THAT volcanoes are, for the most part, good. No, they are *incredible.* Volcanoes are capable of doing things that verge on the supernatural.

Let me tell you about a Japanese volcano named Sakurajima, meaning "cherry blossom island." Its name used to make sense, because it once stood alone in Kagoshima Bay. But one day, the volcano decided that isolation didn't suit it. It yearned for an encounter with the mainland. When it realized that the people on the opposing shores wouldn't build a bridge themselves, the volcano decided it couldn't wait any longer. In early January 1914, it rumbled and grumbled, letting the people know to stand back. On January 12, it exploded to life, sending ash punching through the clouds, flinging sparks into the air, and spilling lava out into the bay in what amounted to the largest volcanic paroxysm in Japan during the twentieth century.[7] The eruption continued until May 1915,[8] by which point it had dumped so much volcanic debris into the bay that the island was now a peninsula, connected with the mainland. More than a century later, that bridge, built in a volcanic crucible, still stands. I've walked over it several times.

Volcanoes all over the world make their own magic. Eruptions at the bottom of the oceans grow shimmering cities of glass. Lakes of molten rock pool at the top of the glaciated Mount Michael, a volcano hiding on an island just shy of Antarctica. Vertiginous tree canopies full of wild animals embellish the steep slopes of Arenal, a volcano spiraling into the clouds in Costa Rica. Wine grapes coat the flanks of Etna, the Roof of the Mediterranean, while its peak coughs, simmers, and sizzles as a thunderstorm of its own design dances in the night above. Lava flinging itself out of the sea 600 miles south of the Japanese mainland is, at this very moment, piecing together one of the youngest islands on Earth. The lava of Kawah Ijen, a volcano in Indonesia, glows blue and purple at night as the sulfur inside bursts into flames. And not too long ago, just outside of Tokyo, two lifelong friends scrambled up a frozen throne of flame to reach a gate lingering far above the clouds.

Earth, however, doesn't have a monopoly on volcanoes. Any world that has either some heat trapped over from its explosive birth or a geologic method of making new heat can make volcanoes. Orbiting our Sun are planets and moons that, simply by cooling down, can create eruptions that beggar belief: those where lava breaches the gravity well of the world and shoots off into space; those resembling demonic spiders; those so hot and effusive that they can outshine the glittering stars suspended in the background; those that, with time, build volcanoes so immense that they change the journey of these worlds around the Sun. There is no need to apply a science-fiction sheen to the volcanoes of the solar system. To do so would only dull their brilliance.

And there is more to volcanoes than their scientifically explicable sorcery.

The solar system is no longer the land of gods and monsters, but the breathtaking theater of mathematics, physics, and chemistry. We have torn down the dusty veils of superstition and replaced them with a complex cartography that charts and chronicles the behavior of worlds so that we may better understand them. There are, however, far more conundrums than definitive answers.

Fortunately, volcanoes dig up secrets for us that no other natural process can. The peaks, craters, and chasms form where they form, look the way they look, and erupt the way they erupt—if, of course, they are still erupting—because the planetary engines far below the surface are operating in a specific way. When volcanoes erupt, they bring with them scientific gold: heat, direct from the belly of the beast itself; gases, trapped in crystals; ancient rocks filled with a hadean chemistry. These components don't just tinker with a planet's surface in the present. They are quite literally the ingredients for the recipe that makes that planet in the first place. They clue us in to why one planet has water and an atmosphere while another world doesn't, let us know where continents are ripping themselves apart in order to create a new ocean, and tell us if a planet's surface is made up of broken puzzle pieces whose movements fashion everything that is made on the skin of the world. They take us billions of years into the past so we can find out how planets come together and open a window into the possible futures a planet could have taken. Volcanoes showcase the extreme resilience of life that is far from human. And they sketch out the ways worlds can end, and the ways they can't.

Volcanology isn't just the study of volcanoes. For individuals hoping to understand the planets and moons that volcanoes belong to, volcanoes are the great big X's on a map. Eruptions are the delivery of the buried treasure.

When volcanoes bring death, I won't shy away from it. But this isn't an exploration focused on their occasional darker side. This is about the magic they make and the secrets they unearth. And although there are millions of fiery mountains in the solar system to choose from, the volcanoes or volcanic lands I've chosen to dive into for this book are, I think, the most overzealous and fantastical of them all. They are all exceptional architectural masterworks that each tell us something foundational about Earth, our one and only home, or the alien worlds to which they are attached.

Like the volcanoes they dedicate their lives to researching, no two scientists are the same. They each have their own stories, quirks, and idiosyncrasies. They are the reason we can listen to the language of volcanoes and comprehend it. And each and every time I had a conversation with one of them for this book, I briefly thought back to my PhD days.

I had wanted to become a volcanologist for the majority of my life (a cliché, I know, but it's true). Originally, though, I had stars in my eyes. After my parents gave me an educational picture book on space phenomena, I became fixated on growing up to become whatever scientist it was that studied the lives and deaths of stellar furnaces. I thought it was crazy that these seemingly permanent bulbs in the night sky were, in fact, just as impermanent as everything else. But it soon dawned on me that, having been born a century or so too soon, exceedingly few people who study the stars would get to go into space. I needed something equally extraordinary, but magnitudes more tangible, to spend my energy on.

An avid video gamer ever since I was four years old, virtual domains had a strong influence on the way I saw the world. And when I was 10, *The Legend of Zelda: Ocarina of Time* was released for the Nintendo 64. Inarguably one of the most influential video games of all time, it blew my mind and gave me a huge 3D landscape named Hyrule. At one point, the protagonist, a boy named Link, gets to explore Death Mountain, a volcano. It isn't by any standards realistic: a huge, multi-tiered labyrinth full of threatening beasts, and the lava, which often pours out of the walls, appears to have some semblance of sentience. But who cares? Back then, the escapades of Link within this cauldron of fire made me wonder if there was anywhere like this on Earth. The United Kingdom, where I grew up, isn't known for its volcanic collection, but I quickly discovered that the world was peppered with volcanoes, many of which appeared to be more bonkers than anything you could find in Hyrule.

My mind was made up. Driven by the desire to study Earth's volcanic splendor, encouraged by my parents, bolstered by enthusiastic schoolteachers, and plagued by the sense that most jobs are horrendous, I ended up obtaining a PhD in volcanology with the hopes of becoming one of the volcanologists or planetary scientists I admired.

But halfway through my doctoral studies, it became clear that the process of conducting the scientific research itself wasn't the source of my elation. Telling stories of these astounding acts of geology was what brought me joy. This thought crystallized in my mind when I met up with my parents in Sicily, a reunion coinciding with my birthday back in 2013 and just a few months before I ended up traveling to Japan. While there, I convinced them to go with me to Stromboli, a small island just north of Sicily where a volcano erupts several times a day.

One day, during the early evening, we walked to the foot of the volcano, through fields of flowers and tall grass. As if on cue, the mountain shook, the air rang, and lava shot up into the night, casting shadows with its fire. We all stood there, mouths open, moved by such an incredible sight. At that moment, I felt the exact same sense of wonder I did back when I was a kid, pushing forward on the controller's analog stick and watching Link explore that volcanic dungeon. I wanted everyone to experience that awe too.

Being a science journalist lets me tell people thrilling tales of volcanic magic, those that eclipse the images of death and destruction to which volcanoes have become synonymous. I want you to feel unbridled glee as these stories sink in and an indelible grin flashes across your face as you think: holy crap, that's *crazy*!

And so here we are, at the start of a journey 4.6 billion years in the making—the beginning of a rollercoaster that winds across Earth, from sky to sea, before leaping up into the starry pool of worlds of volcanic fire beyond. I hope you enjoy the ride.

SUPER VOLCANOES

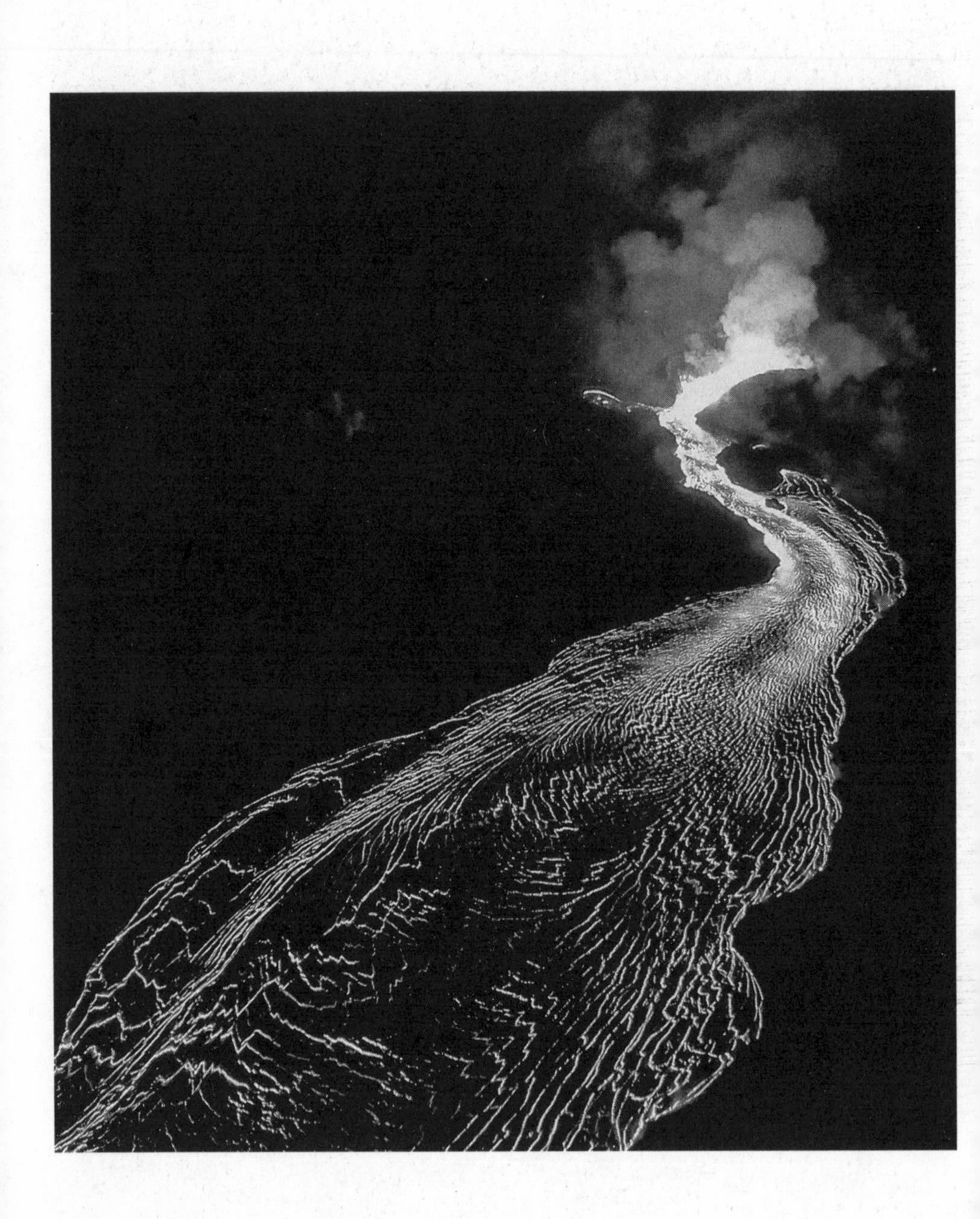

1

THE FOUNTAIN OF FIRE

Toward the end of April 1902, Mount Pelée, a mountainous volcano on the Caribbean island of Martinique, began to seriously stir.[1] Fumaroles—vents expelling hot, noxious volcanic gases—were popping open, the volcano shook from time to time, and fragments of quickly cooling lava showered its flanks. A series of larger explosions sprinkled ash onto the surroundings, contaminating waterways. Animals began to die after inhaling too much volcanic dust. On May 5, a muddy flow of wet volcanic matter crashed through Pelée's crater wall, careened through a sugar-processing plant, and killed nearly two dozen people inside before splashing down into the sea, creating a 10-foot-high tsunami that flooded parts of the city.

Things continued to escalate until, finally, on the morning of May 8, the volcano exploded. Avalanches of superheated volcanic debris screeched down its slopes and quickly engulfed the city of Saint-Pierre, scorching and annihilating all it touched, including boats docked in its harbor. Around 30,000 people were killed in an agonizing heartbeat. There were only a handful of survivors,[2] including a prisoner in the city's deepest, darkest cell, and a shoemaker who lived on the edge of town but was still badly burned as the air around him incandesced.

News of the eruption's lethality quickly ricocheted around the world. That same spring, the US government sent five scientists to investigate the aftermath, among them a 31-year-old Harvard University geologist by the name of Thomas Jaggar. At his arrival and initial exploration of the ruins of Saint-Pierre, the air would have reeked of the stench of death and burned skin emanating from the charred and contorted corpses of thousands of men, women, and children scattered among the stony wreckage of the city. Volcanic ash and boulders lingered in the streets between shattered architecture. The May 8 eruption took place on the Feast of the Ascension, a Christian celebration that commemorates Jesus's rise into heaven. The city's formerly packed churches, whose spires once stood proud and high, were now collections of enflamed rubble. Jaggar stood in what was once dubbed the "Paris of the West Indies"—now nothing more than cinders and bones.

The terror that emerged that day cemented itself in Jaggar's mind after he met two survivors[3] of the eruption at a hospital on Barbados: Clara King, a family nurse, and 14-year-old Margaret Stokes, her protectee. Both had been on a steamship in Saint-Pierre's harbor that grim day, and both suffered injuries; the young girl, in particular, had severe burns all over her body. King recounted the eruption to Jaggar, telling him how enormous rolls of thunderous noise gave way to a hot, dark cloud that frightened her so much she urged death to take her. It was this tale that convinced him to dedicate his research at Harvard to the embryonic science of volcanology. In his autobiography,[4] he wrote that "the killing of thousands of persons by subterranean machinery totally unknown to geologists . . . was worthy of a life work."

In 1906, Jaggar became the head of the Massachusetts Institute of Technology's geology department. On April 7 that year, Italy's Vesuvius—the very same volcano that infamously destroyed and buried Pompeii in the year 79—erupted. Arriving as quickly as he could manage, a few days after the eruption, he made his way to the summit where he spied a broken, bruised, but still standing building that turned out to be the Vesuvius Observatory. Although it was opened in 1848 as a scientific research outpost for the study of magnetic and meteorological phenomena, it soon evolved to become the world's first volcanic observatory. When Jaggar happened upon it, the first seeds of an American equivalent were sown.

Expeditions, experiences, and events in subsequent years, from seeing spectacular eruptions in Alaska's Aleutian Islands to hearing about death-dealing earthquakes in Italy, only served to magnify his enthusiasm for a multipurpose geoscientific observatory of his own. Soon, various philanthropic entities offered their support. A sizable donation from the trustees of the Boston-based Edward and Caroline A.R. Whitney Estate was given to MIT in 1909 in order to construct one, but they insisted the observatory had to be in the vicinity of Boston.

Later that year, while voyaging to Japan with his wife, Helen, they stopped en route to explore the Hawaiian Islands. The islands were born over millions of years as lava emerged from the seafloor, rose above the waves, and formed castles. Many of its volcanoes are now dead, but some, like Mauna Loa and Kīlauea, remain active, their insides rumbling with magma.

But for Jaggar, Kīlauea, on the southeastern side of the Island of Hawai'i, had a unique draw. The couple lodged at its summit, in the Volcano House hotel, 4,000 feet above sea level.[5] Each of the three days he stayed there, Jaggar walked into and across the three-mile-long geologic bowl, or *caldera*, atop Kīlauea to reach a shimmering and spitting pit of molten rock named Halema'uma'u in the caldera's southwestern section. Every day, he saw rocky rafts of volcanic crust twist and turn, sinking beneath the waves of a pool of fire. The air was hot and tainted with an acrid, sulfurous breath. He was transfixed by the lake of lava, a window into the underworld fit for America's first volcano observatory.

It's easy to see why such stunning volcanic magic has inspired so many tales of deific mischief and mayhem among Native Hawaiians, who have lived in the archipelago for around 900 years. Pele, often thought of as the goddess of lava, is also the lava incarnate.[6] Many Native Hawaiians consider her to be their ancestor. Like many origin stories, the accounts of her origins differ, but they generally follow the same beats: Pele's family were traveling from island to island in the South Pacific, hoping to find a home for Pele's flames. A hole was dug in each island, each a temporary housing for the fires. Eventually, the Island of Hawai'i proved to be the best match, with Kīlauea's Halema'uma'u ultimately becoming her place of residence.

By all accounts, Pele is a temperamental goddess. She can be anything from benevolent to vengeful, but she is frequently venerated by many Hawaiians, even when her lava destroys homes and property before cooling down and making new land. Her molten outpourings, both creative and destructive, will go where they will, with anything artificial in her way simply becoming an offering.[7]

It is impossible for volcanologists from overseas to arrive in a distant, culturally distinct part of the world and be able to figure out the nuances of a particular volcano themselves. Local knowledge is always vital. Spiritual beliefs and the scientific method are two very different things, but nearly a millennium of experience living with Hawaiian volcanoes is something that Western scientists cannot afford to ignore. Kīlauea had been erupting long before Jaggar arrived in 1909, and as the decades went by, scientists digging into the earth found evidence for plenty of ancient eruptions locked up in buried rock layers. But the oral traditions of Native Hawaiians mean that descriptions of far older Kīlauean eruptions actually witnessed by people have also been preserved—a vital source of information on the behavior of lava that cannot always be gleaned from old rocks alone.

But when Jaggar first set eyes on the lava lake at Halema'uma'u in 1909, the scientific techniques that could decode the physics and chemistry making Pele's powers possible simply didn't exist. Lava lakes are essentially the tops of the magma reservoirs that are normally trapped miles belowground. Despite what movies have hoodwinked you into thinking, lava lakes are also exceedingly rare. They have appeared and disappeared in volcanoes all over the world. Today, you can count the total number of long-lived, active lava lakes on two hands. Many are also inaccessible, with sources of danger being both environmental (a deep crater with steep slopes) or human (they sit in the middle of a conflict zone).

The world-beating accessibility of Halemaʻumaʻu was a game-changer in Jaggar's mind. His unrelenting nature, and the local support of Honolulu businessman Lorrin Thurston (among others), ultimately meant that Kīlauea, not Boston, was to be the site of his geologic observatory.

Frank Perret, an American engineer whom Jaggar first met at the Vesuvius Observatory in 1906, arrived in Hawaiʻi in the summer of 1911 at Jaggar's behest and quickly proved himself to be instrumental,[8] making detailed observations of the lava lake and setting up a cable across it from which scientific instruments could descend.

Experiments on the lava lake in the early 1910s aimed simply to determine its most rudimentary physical properties. Sure, it looked hot, but how hot? Ordinary thermometers wouldn't survive entry, so bespoke heat-resistant 10-foot-long temperature gauges designed in Baltimore by Leeds and Northrup were sent over to Kīlauea. Two thermometers—electric pyrometers, as they were technically known—were quickly eaten up without any readings being made. Perret, along with E. S. Shepherd of the Geophysical Laboratory of the Carnegie Institution in Washington, DC, were struggling. With no funds available to hire any additional help, Thurston recruited his own family to help operate the cable rig,[9] a hardy mechanical contraption that was nevertheless quickly succumbing to the lava lake's extreme temperatures, corrosive acidity, and changing dimensions.

Their third attempt, in July 1911, was successful,[10] marking the very first time the temperature of molten lava was directly measured. Before the successful pyrometer was also consumed by the lake, it sent back an electrical jolt corresponding to a temperature of 1,832 degrees Fahrenheit. Such a temperature is difficult to imagine, so think of it this way: that's 10 times hotter than the water boiling in your kettle.

Jaggar, who was absent for these experiments, arrived back in the archipelago a few months afterward and later described the machinations of the lava lake in great detail in a series of academic papers. One, published in 1917,[11] emphasizes the unfathomably fiery nature of Halemaʻumaʻu, describing "puffing vents where banners of flame play above . . . cones as natural blow-pipes of burning sulphur and hydrogen," adding that "these vents will fuse steel."

It has no fixed opening date, but many consider 1912 to be the year the geologic observatory Jaggar had long dreamed about was founded. In February, prisoners of the Territory of Hawaiʻi (it wouldn't become a state until 1959) were used to dig the foundations of a volcano observatory post in Kīlauea's old, rigid volcanic rock. In May, MIT appointed Jaggar the director of the Hawaiian Volcano Observatory (HVO), and in July, Thurston started paying Jaggar's salary.

The Hawaiian Volcano Research Association, set up by Hawaiian residents in 1911, was a private source of financial support for HVO for several decades.[12] They chose a mantra inspired by Jaggar's guiding motivation—to understand volcanoes so that future disasters, like the apocalypse at Mount Pelée, could be avoided: *Ne plus haustae aut obrutae urbes*—"No more shall the cities be destroyed."

It is a mission statement that the US Geological Survey, which became the permanent administrator of the HVO in 1947, abides by today. Founded by an Act of Congress in 1879 and tasked with mapping out the nation's valuable mineral resources and public lands, it soon became the NASA of terra firma, with scientists from all walks of life devoting careers to untangling the mechanics of glaciers, rivers, mountains, oceans, earthquakes, and volcanoes—partly to better understand America's natural wonders, and partly to ensure that lives are protected when the next geologic disaster rears its head.

THE CENTURY SINCE its founding has been an astoundingly eye-opening era for the HVO as, with the help of researchers from the

Aloha State's universities, it has kept watch over the archipelago's various volcanoes. Although it has moved its location a few times, orbiting around Kīlauea's summit, HVO remained up on high, staring into Halemaʻumaʻu's furnace as it ebbed and flowed.

Kīlauea is a profoundly active volcano, living up to its name: "spewing" being one translation. But 1983 was a landmark year[13] even by its own standards. On January 3, along a rift zone on its eastern flanks, fissures in the ground opened up, erupting gooey lava like it was going out of fashion. By June, many of the fissures had become inactive, and the lava was flowing out of a single vent. This singular focus of molten matter produced a spectacular, erratic array of lava fountains that, over the course of the next three years, created a large volcanic cone by the name of Puʻu ʻŌʻō, rising 835 feet above the ground, a basilica born of flame.

Despite proving itself a proficient architect, the volcano simply kept erupting. And erupting. And erupting. Months went by. Then years. Then decades. Throughout, HVO staff continued to watch, learn, and understand. Lava flowed. On occasion, it flowed into towns, steamrolling through people's houses at less than a third of a mile per hour. People could easily outrun it. Infrastructure rooted in place was slowly subsumed by thick, dense, scorching slugs of fire.

In 2008, not willing to let the eruptions at and around Puʻu ʻŌʻō occupy the entirety of the limelight, a persistent pool of lava returned to Halemaʻumaʻu, a welcome return after it disappeared during a fitful, explosive-laden eruption in 1924. Despite the destructive nature of some of the lava flows crawling out of fissures on its slopes, scientists, visitors, and many Hawaiians were enthralled by Kīlauea, a volcano that could erupt from its peak and along its flanks simultaneously. And in 2018, the magmatic mountain appeared to be unstoppable.

IN 2015, Christina Neal became the scientist-in-charge at HVO. Her volcanological tour of duty with the US Geological Survey, though, began in 1983, the same year Kīlauea started its seemingly endless eruption. She ventured to and studied volcanoes all over the world, from those in the Galápagos to the tall volcanic beasts of Alaska and Russia's icy Kamchatka Peninsula. Each volcano has its own identity, even if they share some similarities. But in one very crucial aspect, Kīlauea was unlike any other. Many volcanoes, especially the volatile mountains in Alaska, could only be seen from afar. HVO sits right on the summit of Kīlauea, making studying it an altogether more personal experience. "You are living on it. You are breathing air from it. You are walking up to the lava and almost touching it," Neal tells me. "The proximity to you as an individual, and as a scientist, is incomparable."

Although keeping an eye on multiple Hawaiian volcanoes, Neal was cautiously monitoring Mauna Loa, the massive volcano up the road from Kīlauea. If the latter was the life and soul of the party, the former was enjoying a prolonged phase as a wallflower. Several of its past eruptions had been destructive and at times looked like they were about to destroy the city of Hilo. But since Mauna Loa's last eruption in 1984, it had been ominously quiet. What was it up to?

In 2008, Halemaʻumaʻu's lava lake had returned after a spell away, and in March 2018, Hawaiʻi was celebrating its 10th anniversary. "Overall, the (lava) lake for the past several years has been very steady," an HVO geologist told the *Hawaii Tribune-Herald*.[14] "It's not showing any signs of diminishing."

But as the days went by, something within Kīlauea began to stir. Instruments at Puʻu ʻŌʻō showed the ground around it changing shape. The magma pocket below was accumulating pressure, like someone blowing up a bouncy castle inside a house. The pressure spiked, pushing up magma and causing Puʻu ʻŌʻō's lava pond to rise. It looked as if there might be another outbreak somewhere in the East Rift Zone, or ERZ. On April 17, 2018, the HVO issued an alert, informing the public of this possibility.

Something was moving about beneath Halemaʻumaʻu too. The lava lake was ascending so quickly that it looked like it could overflow onto the lake's shores. On April 24, another alert was issued to that effect. Two days later, a large overflow painted old ground with swatches of molten rock. This was curious, but nothing hugely concerning. There wasn't any sign the spill would be significant enough to begin to venture far beyond Halema'uma'u's crater and coat the far greater expanse of Kīlauea's caldera. Even then, no one, save for scientists and guests of Volcano House, lived around or near the summit. As past eruptive activity clearly demonstrated, the real danger lay further down the flanks.

On April 30, Pu'u 'Ō'ō's crater floor collapsed, and its pool of lava rapidly disappeared from view. This had happened before, so on its own it was no cause for concern. But this time it was accompanied by a series of earthquakes marching down toward the Lower East Rift Zone, or LERZ—a site home to dozens of eruptions in the past few thousand years. That, Neal says, was when HVO staff expected a departure from the recent past. Those earthquakes were the sounds of magma tunneling through solid rock. Something new was coming—an eruption in a different place, but what would it be like? How long would it last for? Days? Weeks? Months, or even years?

A century of scientific observations, and centuries more by Native Hawaiians, may sound like a considerable tapestry of time in which to find hints as to how future eruptions would behave. But volcanoes live far beyond human lifetimes. They obey the geologic gods of time, where moments are measured not in years, but millennia. Volcanologists couldn't simply track Pu'u 'Ō'ō's convulsions and determine when the rhythm would change. All they could do was watch, listen, and wait.

You might think that the atmosphere at HVO at the time would have been febrile. But for the most part, there was "a significant amount of seriousness—that's the word that just leapt to mind," says Neal. "Of course, there was a level of excitement and interest, scientifically," she adds, because this big change in Kīlauea's behavior was a great opportunity to learn more about it. But the staff knew that every observation, and every interpretation of each observation, had consequences for the people living in the volcano's shadow. Their lives and homes, once again, were at stake. Not knowing where the magma moving about underground was going to erupt was disquieting.

On May 1, HVO warned the public of an eruption somewhere in the LERZ. Tensions rose. On May 2, small cracks opened in and around Leilani Estates, a residential area of about 1,500 people in the LERZ. The lava lake at Halemaʻumaʻu began to sink. On the morning of May 3, a magnitude 5.0 earthquake shook Puʻu ʻŌʻō, causing the bottom of its crater to sink even further. Later that day, at around 5:00 p.m. local time, a crack in Leilani Estates transformed into an infernal scar in the earth as lava began to ooze out, quickly setting nearby trees ablaze. Gray smoke rose from the fires into the sky.

The cries of a new eruption were heard around the world.

ROUGHLY 2,500 MILES away in Portland, Wendy Stovall was putting her son to sleep. As the deputy scientist-in-charge at the Yellowstone Volcano Observatory, she specialized in Yellowstone volcano's comings and goings. But Stovall, like many of her colleagues at the US Geological Survey, had a diverse background. She happened to have studied Kīlauea for her doctoral degree and, like many volcanologists, she was paying close attention to its exploits from afar. "We all knew *something* was going to occur," she says.

That night, she got a text from Mike Poland, the scientist-in-charge at the Yellowstone Volcano Observatory, saying that a fissure had opened up in Leilani Estates. Here we go.

She hastily put her son to bed, left his room, and paused for a moment—the calm before the storm. A couple of minutes later, she found herself on the media response team. Questions from all kinds of outlets were flying in faster than she could react. The *Los Angeles Times* wanted to chat right that very moment. The country wanted immediate answers: What was happening in Hawai'i?

Every two years, the Yellowstone Volcano Observatory invites all the universities and institutions that participate in its work to an in-person meeting. In 2018, it was to take place on May 7 and 8 at Mammoth Hot Springs in Wyoming. Stovall recalls spending much of it outside the meeting rooms, talking to the media about Kīlauea as snow-covered hot springs effervesced outside the window. "It was a crazy time," she says.

By the time the meeting had ended, chaos reigned. Just five days after the first fissure appeared, 13 more had ripped open the land in and around Leilani Estates—in people's yards, across roads, right in the middle of neighborhoods. Fissure 8, opening near the edge of Luana Street, began shooting a fountain of lava into the air while smoldering volcanic putty built a small cone around it. The stench of sulfurous compounds hung in the air as lava roared skyward, impersonating the thundering rumble of a jet engine.[15] Some fissures struggled and ran out of juice while additional fissures opened up and began erupting. Newcomer fissure 17 stole the spotlight on May 13 by launching lava bombs—clumps of cooling lava still warm enough to change shape in mid-air—400 feet into the air, accompanied by booming explosions heard miles away. And all these events were documented by geoscientists and promptly relayed to the Hawai'i County Civil Defense and the Federal Emergency Management Agency as they guided those fleeing the fires to safety.

Video of a Ford Mustang being slowly but surely consumed by lava went viral on social media, crystallizing in people's minds the fact that lava will go where it pleases, no matter what stands in its way. Towns around the fissures were ordered to evacuate, but some were quickly allowed back into the danger zone in order to retrieve vital medication and stranded pets. Some saw their houses go up in flames as lava touched their walls. Others received grim news remotely. One evacuee realized that her home was doomed when its security system's motion sensors were tripped[16] as lava began wandering through her hallways.

Just 25 miles away, Kīlauea's summit was making its own pyrotechnics. From May 4 onward, it shook with increased intensity and frequency. Halemaʻumaʻu's lava lake began to sink—fast. By May 10, the lava could no longer be seen, having descended 1,070 feet below the crater floor. Thermal cameras chronicled the stunning vanishing act: a lake of red rafts surrounded by searing yellow boundaries dropped out of sight at breakneck speed, leaving behind a cool palette of magentas and violets. And this disappearance wasn't happening in isolation. It indicated that the magma reservoir beneath Halemaʻumaʻu was draining away. That meant the entire summit caldera, no longer supported by the magmatic foundations below, began to sink like a deflating balloon. The volcano, suffering from indigestion, also belched out lung-scratching volcanic smog, or "vog": a visible haze of volcanic dust, sulfur dioxide, and droplets of sulfuric acid.

Earthquakes shook rocks into the summit's emptying pit, and explosions shot ash thousands of feet into the air. Neal, sitting in her office at HVO headquarters on the caldera rim, had a front row seat to the summit collapse as it began to swallow the land around it. But she spent much of her time looking elsewhere: at her computer as a torrent of data kept arriving at all hours of the day and night; at plots, sketches, and visuals of Kīlauea's mayhem in meetings with her deputies and staff; into the middle distance as scientists in the LERZ called to report the latest. The streams of emails flowing to and from scientists at partner universities were never-ending. Requests for decisions came in thick and fast. She says she felt like the coach on the sidelines of a sports game, constantly giving players instructions while trying to peer over the temporal horizon.

It wasn't long before Neal and her colleagues found themselves under siege. As they sat on the periphery of the giant cauldron, they felt it growl, and the earth began to judder and jolt. Every 36 hours, a magnitude 5.0 temblor caused the entire summit, including the building they were housed in, to violently convulse. It was like working inside a tumble dryer.

Literal cracks began to show. Within the first few days of the eruption, Neal says, "the earthquake activity was getting to the point where the building was breaking around us." By May 10, it was crystal clear that trying to respond to a volcanic disaster in a disintegrating building was untenable. Gathering all of their scientific equipment and IT infrastructure, as well as the invaluable contents of the archives in their cavernous basement, they evacuated their headquarters on May 16 as the moving earth refused to relent. They moved 30 miles away to Hilo, monitoring the eruption as they went.

By this point, Stovall had arrived on scene. She was one of the many scientists from across the country that HVO had asked to fly to Kīlauea to help out with the eruption response. On her first night, she found herself talking to the area's fire chief. Realizing she hadn't yet ventured into the field, he immediately drove her and a friend around the island to a spot where astonishing flows of lava could be seen in all their mercurial majesty. Stovall studied Hawaiian lava fountain eruptions for her PhD, but outside of old photographs and footage, she had never seen one with her own eyes. Finally, the time had come. "Witnessing it and understanding the processes that were occurring in order to create the thing I was seeing—fundamentally understanding those processes—it was rewarding," she explains. "It took my breath away."

Stovall was there primarily to liaise with the media. At one point, she went along with a group of journalists on a National Guard–guided tour into the fissure-scarred realm. As she stared into a lava fountain, the physics of it danced about in her head. A local television reporter, filming via livestream, asked her several questions about it. At first, she supplied him with a mix of oft-repeated US Geological Survey–approved statements and scientific exposition. Noticing she appeared to be distracted, the reporter asked her what she was preoccupied by.

The volume of the volcanic soundtrack playing in the background of her mind was suddenly jacked up. Her eyes lit up. She excitedly waxed lyrical about the remarkable nature of the eruption—the geology, the physics driving it. In that moment, she says, she felt like she was doing the very thing she had been building up to do throughout her entire adult life.

It wasn't just the lava that bewitched Stovall. The increasingly voluminous lava flows caused the air above them to sizzle with heat. Big patches of warm, damp air rose, eventually reaching a height cool enough to allow all that moisture to bunch around ash and precipitate out as vast billowing ghosts called pyrocumulus clouds. Lightning flashed, thunder boomed, and acid rain coated the earth, eating away at any plants the lava hadn't already flattened. As you drove toward the horizon, you could see this lava-made weather long before you could spot lava crawling on the ground below—a sight, says Stovall, whose extraordinary nature is impossible to convey with words.

AS LAVA MEANDERED across Kīlauea's flanks in early May of 2018, earthquake geophysicist Ken Hudnut was 4,800 miles away, perusing through satellite shots of tributaries of molten rock at the headquarters of the US Geological Survey (USGS) in Reston, Virginia.

The USGS had access to plenty of technologically advanced satellites, equipped with sensors that see much of the surface of the planet in high resolution no matter what tried to obscure it. That was great for geoscientists wanting to look at, say, the movement of lava or the development of a new volcanic cone otherwise shielded by a fleet of clouds, plumes of vog, or puffs of volcanic ash. But such penetrative surveillance hardware can also have applications in the world of espionage, so anyone at the USGS wanting to access certain visuals needs security clearance. That's why the room containing all this imagery, accessible to a select few, was affectionately known as the spy shack.

Around that time, it became clear that lava was moving toward

Puna Geothermal Venture. Like other geothermal plants, this one used steam generated by fluids mingling with hot rocks to spin turbines around. You can use water for this, but this plant used pentane, a substance whose eagerness to boil allows it to generate turbine-spinning vapors faster than steamy water alone. Pentane, however, is highly explosive—so as lava began to creep toward the plant, its pentane supplies were quickly rushed off-site.

Danger remained in the form of the deep boreholes used to tap into the hot rocks below. These are normally pressurized, and lava heating up their insides could cause them to explosively pop open.[17] If they did, hydrogen sulfide, a volcanic gas, could be unleashed on the area. At low concentrations, it smells like rotten eggs. Higher concentrations in the same volume of air are not only odorless[18] but can also cause an excess of fluid to accumulate in the lungs, leading to unconsciousness and even death.

The plant's wells had been quenched with cold water in order to lower their pressure, and few if any staff were left on-site. But lava was still advancing toward them, and the thick vegetation around the plant meant that helicopters and satellites were having trouble seeing exactly where it was going. Hudnut, noticing this problem and looking at imagery obtained from the spy shack, had a solution: fire lasers at it. Specifically, he wanted to attach a series of differently angled lasers to a helicopter, fly them over the power plant, and use pulses of their light to pierce through the vegetation like it wasn't there, letting scientists know what the ground there looked like and if it appeared capable of funneling lava to these accidental bombs.

This technology, named light detection and ranging, or LiDAR, is capable of virtual deforestation, something prized by archaeologists wanting to find buried historic treasure hiding beneath thick canopies. Hudnut suspected that the same science could be used to chart Kīlauea's terrain, virtually scything away trees and bushes to spy the cracks, crevasses, and slopes likely to guide lava to new destinations. His supervisor approved of the idea, and he was soon on a plane to Hawai'i.

Meanwhile, things kept escalating. A major summit explosion on May 17 created a plume of ash 30,000 feet high, a little shy of the cruising altitude of passenger jets. The summit continued to deflate. Down in the LERZ, rivers of lava were bifurcating, and on the evening of May 19, one tributary crossed a highway and started pouring into the Pacific Ocean. Giant plumes of white ephemera began bubbling up from the coastline, a mixture of water vapor, hydrochloric acid, and tiny specks of hot volcanic matter that had been quenched into fine shards of glass. Although it diffuses as it rises, this so-called lava haze, or "laze," can be deadly to anyone standing too close.

The first major injury of the eruption was reported around this time: a man up on a balcony at his house—one he shouldn't have been on by that stage—was unable to dodge a bomb of molten rock that was launched out of a nearby fissure. You may think it's the excessive heat you should fear, and that's fair enough: no one wants a substance ten times hotter than boiling water splashing over them. But the real problem is the density of the lava. This isn't syrup we are talking about here, but molten rock—and volcanic rocks tend to be pretty heavy. I imagine the scalding temperature of the projectile was deeply unpleasant, but having something that weighs as much as a refrigerator flying into you at hundreds of miles per hour was probably worse. The man survived, but only because the lava clump didn't shoot through any of his vital organs. Instead, it smashed through his shin, shattering it into pieces.

A few days later and the eruption produced another act of sorcery. Near some of the fissures in Leilani Estates, people began seeing blue flames trip the light fantastic, whirling and twirling in the night. One journalist covering the eruption, Maddie Stone, described the deep blue fires with unassailable magnificence for *Gizmodo*,[19] writing that the scene was reminiscent of "bioluminescent waves lapping against the shores of hell." Several things can cause blue fire, including the ignition of volcanic sulfur, which in the past has cast an otherworldly hue on the summit. In this case, the lava was overrunning buried vegetation, cooking it and causing several organic compounds to leak out. One of these was natural gas—mostly methane—which, when rising up to meet the lava, burst into blue embers, not that different from the pyrophilic gas lighting up your kitchen stove. Like all of the eruption's magic tricks, this one also brought with it an element of danger. Pockets of methane gas could build up before being set alight, which could cause an explosion not that different from that of a buried hand grenade—a blast very capable of lopping off anyone's legs if they stood too close.

By the end of May, fissure 8 really upped its game, creating and sustaining fountains 250 feet high. Lava coming from it was moving in multiple directions, making its future migration more unpredictable. "I did feel like, in those early days, that it was possible that somebody was going to get trapped and possibly killed," says Neal. Scientists constantly updated the county civil defense staff and hoped people would remain unharmed, but in the LERZ they often saw people in places they should have long since abandoned to the fires.

Stovall's moments of joy at seeing Kīlauea's grandeur in person were tempered by times of grief. She had been working around the clock, with little sleep, for two weeks straight, joining and sometimes facilitating community meetings with dozens of people who had seen their homes and livelihoods melt away before their eyes. One day, she was stopped in the street by a woman in tears. She wanted to know if her house was still standing but couldn't bring herself to go and check. Stovall did it for her. The lava hadn't yet reached her house, she told her, but it was only a matter of time.

An especially difficult moment was relayed to me over our Zoom call. Stovall's tale was punctuated by several long pauses, and her voice briefly wavered a couple of times.

At one point during her tenure, she went up to the second floor of someone's house to see a lava fountain right outside the window. Later that night, the house was obliterated, and the next day, lava was quickly flooding the area. A community meeting was held to advise the public, and a crowd of people approached her, asking if their homes in Kapoho and the Vacationlands neighborhoods on the coast would be consumed by the lava. Topographic maps of the region suggested the lava would likely continue to move north, avoiding their homes. "You guys don't have anything to worry about," she told them.

That night, after the meeting, the lava followed the road to the east and began to plough right into their houses. Residents were jolted awake by the sound of sirens. A voice over a loudspeaker kept saying "evacuate" over and over.

"When I woke up the next morning, and I'd seen the progress of the lava, I felt so horrible," Stovall says. She learned a hard truth that day. When dealing with a dangerous eruption, "you can't come from a place of hope. You have to come from a place of facts," she tells me. "You can't get your heart into it when you're trying to convey hazards information to a potentially vulnerable population." She feels like she did them a disservice by not being clearer and saying, we aren't really sure, it looks like it might avoid your neighborhoods, but it could take a turn. "And I wish that I could tell those people that I'm sorry."

Two weeks in Kīlauea's shadow was sufficient time in the field for Stovall. "That was enough," she says. "It was traumatic."

By the start of June, fissure 8 continued to flare as around 3,500 cubic feet of lava fell into the ocean every second, which is enough to fill up every room inside two average American houses in a minute.

SATELLITE IMAGES OF lava engulfing plants and property alike really provided a breathtaking sense of the scale of the eruption, but they sometimes felt coldly removed from the devastation on the ground. Footage filmed on television cameras and smartphones brought viewers into the heart of the chaos, but any wider perspective was lost. But by this point, the transformation of this corner of the Island of Hawai'i was also being seen through an altogether more cinematic lens: those attached to unmanned aircraft systems or UASs, better known as drones. Orbiting around fountains of lava or hovering over fissures as they tore the ground to pieces, drones gave viewers both perspective and a frightening intimacy, a window to the most visually resplendent volcanic house of horrors in living memory. You could be thousands of miles away from the action and still feel the heat haze on your face through the footage.

Before the eruption began, Angie Diefenbach, a geologist at the US Geological Survey, seemed to be fighting a losing battle. "I'd been wanting to use drones for a long time with our domestic program with the [USGS], but management never really saw the utility of it," she says. As drone technology improved over the years, she found herself using them with overseas colleagues in the field. She got her pilot's license and procured some drones for the USGS, hoping one day to use them on American soil.

By the time fissures started turning Leilani Estates into a bonfire, she was still the only person in the USGS Volcano Science Center that had a drone pilot license. She saw her chance. Drones could go where scientists could not, flitting right above active lava flows and fissures, providing vital scientific data to the HVO. The University of Hawai'i at Hilo had also deployed its own drones right at the start of the eruption, paving the way for the USGS to do the same on a larger scale. Convinced by her case, Diefenbach was sent to Kīlauea with eight drones: two hexacopters, four quadcopters, and two fixed-wing aircraft, piloted by a team of five, eventually expanding to 36 pilots recruited from across the Department of the Interior.

Not only was the footage of slithering, racing lava jaw-dropping, but also the data it obtained—the dimensions and characteristics of the lava, how it was flowing, where it was going, what noxious gases it was emitting—gave volcanologists a game-changing new way to study the eruption. It provided them with the first bird's-eye view of the increasingly open maw at the summit, whose occasional ashy plumes proved too risky for helicopters. "We were always very careful and super cautious, but we did push them to their limits, I think," says Diefenbach. "That's the great thing about drones. You're not risking any human lives when you're flying them."

They were also instrumental in shielding civilians, and not just because the county civil defense staff could track, in real time, volcanic threats as they emerged. One night, as drone footage was being used to map out lava flows, the team was alerted to the fact that a man was stranded in his house and had become almost entirely surrounded by lava. One of the drones flew to his location before guiding him through the night along a path the lava had yet to cross. If it weren't for the drones, he might have perished.

It was the "perfect application of a perfect tool," says Neal—the most peaceful deployment of drones in American history, saving lives rather than ending them.

As the lava gobbled up neighborhoods, the drone pilots also offered a sense of closure or comfort to evacuees, with those wanting to know if their homes still stood being shown either the harsh or happy reality on the pilots' tablet computers. "I think it brought them a lot of peace, in some ways, that we could provide that information on demand and in real time to them," says Diefenbach. That quickly obtained and easily accessible footage was also used by evacuees to get financial relief—far sooner than they would have received aid if a ground crew had to investigate and document residential damage on foot.

While camera-wielding drones were creating 3D topographic models of the lava-strewn land, LiDAR-equipped helicopters zipped back and forth, firing lasers and charting every crack and crevasse. By day, the aircraft zipped across the summit and the LERZ, gathering data. By night, Hudnut and his colleagues stayed up into the early hours, transforming gigabytes of data into astoundingly accurate, three-dimensional digital models of the volcano as its summit sank and its eastern flanks were reshaped by the magmatic conflagration. It was an evolving map of Kīlauea anyone could view from their desktop computer.

Not everything about the scientific operation was high-tech. Umbrellas were used to stop certain instruments getting wet in the rain, while shovels were thrust into lava flows as scientists stole fresh samples of molten rock to study. Astoundingly simple heists like this allowed scientists to realize that around a month into the eruption, the lava had gotten hotter—perhaps exceeding 2,200 degrees Fahrenheit—and runnier. The early stage of the eruption expelled the underground cache's magmatic leftovers from a previous ebullition. Now it was tapping deeper magma that was gushing out over the LERZ, a hot mess that could sweep across the land with greater ease.

On June 2, lava heading seaward made an incursion into the two-acre Green Lake, the largest freshwater body in the archipelago. In just a few hours, after fizzing vigorously under a sheet of laze, the entire lake had disappeared, replaced by lava that seemed unaffected by the cooling effect of the water.

That lava went on to destroy hundreds of houses along the coast before falling into the Pacific. So much molten rock was being thrown into the sea at this point that it was able to stack atop itself along the shore. By mid-June, it built a brand-new, 250-acre coastline of frozen-over lava.

IT WOULD BE TEMPTING to think that for all the advances in science and technology coming about since HVO was founded, researchers would have discovered a way to stop or at least divert the lava flows. But history tells us that, for all our academic adroitness, humanity has yet to become a match for magma.

You could try dumping a lot of water on the lava to freeze it. The residents of Iceland's Heimaey Island did just that in 1973 when the eruption of Eldfell volcano threatened to close their harbor.[20] Around 1.5 billion gallons of seawater slowed the flow enough to win the day, but only because this lava was already very slow moving. By comparison, Kīlauea's lava was galloping along, with much of it not simply lingering by the sea but spreading out across a huge area further inland.

You probably can't bomb a lava flow into submission. America would know, because it has already tried it—twice.[21] In 1935, lava from Mauna Loa was flowing toward Hilo at a mile per day. The lava wasn't just flowing aboveground. When its outer edges froze, it formed its own channels of molten rock; when its roof froze, it snuck about below the surface in rocky tubes, the insulation of heat allowing it to remain molten many miles from its source vent.

The head of the HVO, our friend Thomas Jaggar, called on the US Army Air Corps to help. Just after Christmas Day, 10 Keystone B-3 and B-4 biplane bombers dropped 20 bombs, packed with 355 pounds of TNT each, on the lava, hoping to hit some of its channels or cave in its tunnels that were funneling the lava toward the town. Hilo was spared when the lava stopped flowing on January 2, 1936. Jaggar claimed victory, but other experts, including the pilots who dropped the bombs, thought it to be a coincidence. In 1942, bombs were unleashed on a hostile lava flow for a second time—again, these efforts were in vain.

That's not to say bombs planted on foot couldn't technically work. During the 1991 to 1993 eruption of Mount Etna, on the Italian island of Sicily, attempts to divert lava away from the town of Zafferana Etnea with barriers had failed. Explosives designed to blast open new diversionary channels also failed repeatedly until one attempt finally ended up succeeding, saving the town.[22]

There was no chance that explosives were going to be used in 2018 on Kīlauea. The lava was too fast, too runny, too unpredictable to try and get it to change direction. Engineers wouldn't have time for a lengthy process of trial and error. Even if the lava was deflected, there was no telling what it might then run into.

Early on, Hudnut wondered if any sort of diversionary process or any serious attempt to forecast exactly where the lava would go would be effective. Conversations with local Hawaiians underlined just "how naïve that was," he says. Lava does not flow like water. It forms its own channels, tunnels, dams, and waterfalls. It rapidly carves out its own topography. All scientists could really do was watch things unfold, like a theatrical audience engrossed by an enrapturing multiact play.

By mid-June, while some fissures huffed and puffed noxious gases, fissure 8 became the focal point of the LERZ eruption, feeding what was now a channel of seaward-bound lava while constructing a 180-foot-high cone around itself. The fury and magmatic productivity of this particular abyss was a staggering display of volcanic shock and awe. Hudnut, one of the many scientists to stand near it wearing flame-resistant suits and specialized breathing equipment, tells me it was a transcendent, life-changing experience. When it was able to produce its own thunderclouds and acid rain, it was also a somewhat unsettling one. Here was the planet forging a cathedral of lava, surrounded by a sea of liquefied crystals—a sight as beguiling as it was belligerent.

BY JUNE 15, Halemaʻumaʻu was more than half a mile across and 1,210 feet deep, more than enough to eat several skyscrapers. HVO's old building was empty, and tourist occupants of the Volcano House hotel had long gone. In their place was Don Swanson, scientist emeritus at HVO, who took up residence in the hotel as the breach in the world greedily expanded.

He didn't go to the LERZ to see the destructive effects of the lava. "I was at the Mount St. Helens eruption 40 years ago, and I got kind of burned-out there by all the devastation that that eruption caused, the social turmoil and the loss of life," he says. That loss included his colleague and friend, David Johnston, a US Geological Survey volcanologist who was one of 57 people killed by the volcano on May 18 when it suddenly erupted out of its side with a singular, shocking ferocity.[23]

That ruination was enough for one lifetime. "I didn't really want to see that again in 2018," Swanson says. Instead, he gave himself a front-row seat atop the screeching, growling crater. The collapse, he recalls, wasn't continuous. It was erratic, with the caldera slumping at its center before suddenly and dramatically dropping down as if the gods themselves had hit it with a hammer. The magnitude of each collapse event varied. Was the next one going to eat up six feet, or 600 feet? Swanson was unperturbed. He was content playing Russian roulette with the sinking summit, taking in the roars of the cave-ins and the turbulent pillars of ash, at peace.

In the same hotel, working out of the dining room, was Kyle Anderson, a geophysicist at the USGS Volcano Science Center. He spent much of the eruption monitoring the summit too—and he could hardly believe how quick it was sinking. Like Swanson, he and his colleagues noticed that the collapse was taking place in increments.

As the magma below zipped elsewhere, the rocky cap above drooped downward. Magma continued to drain, and the pressure it exerted on the cap began to decline. After enough drainage, the friction keeping the circular caldera floor attached to its walls was defeated by gravity. The entire floor shot down, like a piston. It thudded down with such force that the ground around the caldera bounced upward and outward.

Scientists such as Anderson also spotted another reaction to the piston firing: earthquakes slithered downslope, and a surge of lava gushed out of the fissures in the LERZ. It took a lot of careful observations at both the summit and along the flank's fissures over many weeks, but it became clear that the summit collapse and the lava flows were part of a volcanic waltz.[24] As the piston drove downward, it squeezed the draining magma downslope, a bit like someone jumping on a tube of toothpaste 25 miles long.

"I often say Kīlauea is like an organism," says Swanson. What happens to one part affects the other—something made explicitly clear in 2018.

Those piston-powered surges of lava continued to concoct volcanic wizardry downslope. Video footage emerged of coils of lava twirling about above the fissure 8 lava flow. These blazing swirls reached the treetops, spinning fast enough to fling lava across the landscape like a frenzied food blender full of holes. The heat radiating off the lava had destabilized the winds in the region, allowing for vortices to form. Although they were officially called fire whirls, scientists unofficially gave these brilliant corkscrews the denominations they deserved: "lavanadoes."

By mid-July, fissure 8 reigned supreme among the other fissures, most of which had given up the ghost. This corner of the Island of Hawai'i was now a patchwork of black steaming rock adorned with ephemeral veils of laze. The occasional house or patch of green stood incongruously among the desolation. Tourists, drawn to the radiance of the waterfalls of lava falling into the ocean, wobbled about in boats of organized tours. They felt safe from the destruction happening far from their vessels. On July 16, that illusion shattered as a basketball-sized lava bomb shot out of the sea.[25] The bomb, along with hundreds of smaller pellets, crashed through the roof of one of these tour boats like a shotgun blast. Although terrified, the passengers survived, but more than a dozen who suffered burns were sent to Hilo Medical Center, while another with a major leg wound was flown to a hospital.

THE FIRST TIME Emily Mason stood next to fissure 8 and its channel of molten rock was also the first time she had seen lava right in front of her. "It was moving in this torrent—as river rapids. I never thought I would see lava like that," she says. "Every now and again, the crust would almost subduct underneath itself, and you'd just see these flashes of the red underneath."

The doctoral student of volcanology from the University of Cambridge was part of a multi-university team of five volcanologists based in the United Kingdom—"all women, which is a bit unusual." They were asked to fly to Kīlauea to do something they knew plenty about but which the US Geological Survey didn't have the time or capacity to study: the toxic metals and substances sprayed into the air by the eruption, such as lead and arsenic. Although known to contaminate water supplies and corrupt lungs, it wasn't clear how prolific the problem would be during Kīlauea's grand eruption, which looked to continue for months if not years.

Their arrival on the volcano's eastern slopes was a far cry from images of the Sun-drenched, viridian paradise many imagine when they picture Hawaiʻi. Volcanic debris had choked the land of life. Acid rain fell from time to time, forcing them to wear hazmat-like suits. Massive fumaroles had just opened up in people's gardens, encrusted with sulfur and spitting out pungent gases. For reasons unknown, black chickens were often spotted in areas badly affected by the eruption.

The height of the lava in the channel would sometimes appear to drop, a possible precursor to the arrival of another bank-busting, piston-driven surge of lava. The team was standing close to the channel on one of these occasions. Their instinct told them to run, but a "battle-hardened" colleague at the HVO told them to calmly walk uphill to avoid the oncoming flood. Running on uncertain ground could end disastrously if the roof of the freshly baked earth caved in, sending a scientist into a raging underground river of lava.

The team spent three weeks in July navigating the zone of death, using helicopters and drones to flutter above or around volcanic plumes and investigate their harmful ingredients. They also packed a series of scientific instruments into a back-worn frame called Harry, which was named after the protagonist of the volcano disaster movie *Dante's Peak*. Harry would be taken as close to volcanic fissures as possible, dropped, left alone to study the air for a few hours, then retrieved. He was never lost to the fire, but the ground was so hot that parts of him melted.

On August 4, with their work complete, the team was at the airport waiting to fly home. By that point, 320,000 Olympic-size swimming pools' worth of lava had covered 14 square miles of the island while adding around 900 new acres of land along the shore. Kīlauea's summit had collapsed by 1,640 feet, almost the same height as One World Trade Center in New York. The top of the volcano continued to devour the earth as fissure 8 kept pumping out lava. Around 700 homes had been burned and flattened. The economic damage—jobs lost, tourism down, infrastructure and property destroyed—had amounted to hundreds of millions of dollars.[26] The eruption had comfortably earned the ignominious title of the most destructive in the volcano's history. And on August 3, Kīlauea was still showing no signs of de-escalation.

But as they sat in the airport's departure lounge, Mason and her colleagues received an eyebrow-raising communiqué from their friends at the HVO: everything had, well, stopped.

The output of lava by fissure 8 had abruptly dropped, and seismic activity at the summit had declined precipitously. Sulfur dioxide levels had begun to plummet. Initially thought to be a pause in the eruption, volcanologists maintained an air of caution. But by August 7, the fissure 8 deluge of lava was little more than a dribble. The river of lava stretching to the coast rapidly crusted over. The ground, from the LERZ to the summit, no longer twitched in response to any migrating magma movement below. A bleb of lava reared its head inside fissure 8 at the start of September, before quietly disappearing into the invisible night.

Weeks passed. Then months. Nothing more than a weak incandescence, reminiscent of the last embers of a cigarette left abandoned in an ash tray, could be seen during overflights of fissure 8.

The Smithsonian Institution's Global Volcanism Program declares an eruption to be over if three months without any significant volcanic activity has passed. On December 5, 2018, that milestone had arrived. It wasn't just the eruption sequence that began in April that had run its course. The volcano as a whole, which started an eruption sequence all the way back in 1983, had fallen silent. The last few months had been revealed as a frenetic crescendo of creation and destruction at the end of a 35-year-long symphony.

Then, in July 2019, a green pool was spotted in Halema'uma'u by helicopters flying overhead. Two more soon joined it, quickly merging into a single off-cerulean pond of water, kept acidic and hot by escaping volcanic gas; fumaroles on its flanks spurted out gases at 392 degrees Fahrenheit.[27] It wasn't rainwater: heavy rains during the previous few months were unable to gather in the pit. Halema'uma'u's floor now sat 167 feet below the water table, suggesting the source of the pool was water flowing sideways underground and into the crater, a bowl still hot but not hot enough to vaporize water.

Water kept on flowing, and by June 2020, the pond was 120 feet deep and 300,000 square feet across. Scientists were witnessing the birth of a crater lake where lava once gurgled—the first definitive crater lake in 200 years. It was a shame to lose the lake of molten rock, the huge boon to volcanological research that it was, says Neal. But the air quality up there is magnitudes better now that Halema'uma'u isn't eructing noxious gases. "I don't miss that at all," she says.

At the time, volcanologists considered Kīlauea to be down, but not out. The summit had been seen periodically inflating again, indicating magma was slowly gathering deep below. History suggested that, one day, probably years into the future, lava was likely to return to Halema'uma'u. But it was perfectly possible that it might not. Kīlauea could open up a new vent somewhere else on its slopes, and the summit would remain forever unkindled.

After Kīlauea's eruption had been quenched, several of the scientists who had spent 2018 scattered across the shield volcano were thinking about beginnings and endings.

The former was a little easier to tackle. The grand finale of the 35-year-long eruption started when the flow of lava gurgling into Puʻu ʻŌʻō had dropped off back in April. There was something stuck in the plumbing: perhaps an underground rockfall sealed off a conduit to the cone. Consequently, the flow of magma pumping through the rocky arteries that fed Kīlauea's heart increased. Pressure rose, Halemaʻumaʻu overflowed, and the volcano began inflating. Eventually, the plug below the summit gave way, a new artery had appeared, and the heart of magma drained toward the LERZ.

Not much media attention focuses on why eruptions stop, likely because, for storytellers, they are a bit of an unclimactic disappointment. But this eruption sequence had a truly bizarre ending. It didn't slowly peter out and die over days or weeks. Within a single day, the lava factory was abruptly shut down. It was like being in the middle of a concert during an unexpected power cut, the sound instantly vanishing as the venue falls into darkness. Scientists were puzzled but assumed that the volcano had simply run out of magma to erupt, even though Kīlauea did not give any indication that this was the case.

As everyone ruminated over the eruption's sudden termination, Anderson and his colleagues were crunching the data they obtained. By tracking the descent of the lava lake, and by comparing the shape of the summit before the eruption sequence began to its shape after the sequence ended, they could estimate how much magma had been depleted from Kīlauea's shallow reservoir. In March 2018, this reservoir contained anywhere from 0.6 to 1.7 cubic miles of magma. But by December 2018, only 0.2 cubic miles had been erupted.[28] That means no more than 33 percent of that cache was depleted by the eruption's end. It is possible that just 11 percent of its reserves had been expunged.

Put another way, Kīlauea's three-month-long epoch of land-eating, earth-melting, society-destroying madness may have involved as little as one-tenth of its available magmatic fuel. Its volcanic engine was mostly full by the time its eruptive activity dematerialized. On the one hand, this underscores the transformative power of a relatively minuscule amount of lava. On the other hand, it gives volcanologists a real problem: if Kīlauea had so much left to give, why did it suddenly slam on the brakes?

No one could say for sure. Even educated guesses are accompanied by exaggerated shrugs. "We don't really understand what the magma reservoir is like," says Swanson. The changing shape of the volcano at the surface lets us know whether the reservoir is taking on or losing molten rock. Seismic waves released by earthquakes bounce off the magma cache, giving scientists a vague idea of its dimensions. But short of a huge technological leap, says Swanson, reservoirs will remain fuzzy and puzzling. Back in 1911, Jaggar and his comrades began the long task of dismantling the cryptic subterranean machinery beneath Halema'uma'u. That task is nowhere near complete today, and likely won't be for centuries to come.

The path to volcanic enlightenment may stretch far beyond the horizon, but an abundance of critical knowledge was nevertheless extracted from the eruption. Changes in our understanding of volcanoes are usually incremental. But the wealth of preexisting scientific comprehension, the sheer number of scientists involved, the extensive monitoring effort, and the cutting-edge technological marvels deployed mean that this eruption has engendered a giant leap forward for the science. Entire careers will be occupied with distilling answers to long-standing queries from the vast reservoir of data obtained over those three manic months.

Already, two key revelations—that volcanoes can stop erupting with their tanks still shockingly stocked up, and that a devilish dance can exist between the summit of a volcano and its flanks—have given volcanologists cause to cast a suspicious eye on other volcanoes around the world that they previously thought they understood. Thanks in large part to the outstanding success of Diefenbach's drone deployment, the US Geological Survey now has a dedicated drone program. In 2018, she was the only trained drone pilot with a volcanological education. By 2020, there were seven more. The idea, she says, is that every volcano observatory in America will have a crew of drone pilots with a background in volcanology.

The HVO, currently in Hilo, will keep monitoring the volcano, as will geoscientists at universities around the island and the wider world. The comparatively calm exterior of Kīlauea, and its grumbles at depth, are unquestionably fascinating scenes for scientists to study. But there is no doubt that after 35 years of lava, a total lack of it is jarring. "For most of us, for most of our lives, Kīlauea was just erupting," says Anderson. "Now, it's not."

The scars of desolation remain. The fissure 8 cone still stands, a magmatic memorial to the chaos. "For a while, there was a flag at the top," says Neal, placed by climbers. As the community deliberates on which Hawaiian name it will be adorned with, the recovery operation continues. Roads are being repaved. Facilities are being restored. But some sections of that corner of the island will be uninhabitable for decades to come. Many won't ever return to live in the neighborhoods they once called home. Hearts ache, years later.

Kīlauea's 2018 showstopper exemplified the paradoxical trifecta emblematic of many of Earth's volcanoes: their eruptions can be exhilarating, enlightening, and harrowing all at the same time. Like the self-destruction of Mount St. Helens in 1980, which made deadly explosive volcanism an issue for the continental United States, this profusion of lava "is going to be an iconic eruption," says Swanson. It reminded scientists that volcanoes are still more enigmatic than they are familiar. It reminded the world that volcanic eruptions are both the privilege and price many pay for existing on a living planet whose innards are still burning.

Back when the lava still flowed, Neal lost sleep worrying about the many dangerous turns the eruption could take. When I last spoke to her, in June 2020, it was during her final days at the helm of HVO. She sounded buoyant and unburdened. Soon, she'd be heading back to Alaska and its mountainous, snow-capped volcanoes, the end of a 35-year-long eruption serving as an appropriate time to seek a change of scenery. And what a legacy to leave behind: after three months of turmoil—featuring hundreds of explosions, dozens of fissures, countless lava flows, a deluge of bombs, blocks and bank-bursting floods, lavanadoes, blue fires, acid rain, and corrosive clouds—the most remarkable fact about the 2018 eruption is that no lives were lost. For their efforts, Neal and her indefatigable HVO team were finalists (not, unbelievably, winners) in the Science and Environment category of the 2019 Samuel J. Heyman Service to America Medals, the Oscars of the federal government.

Neal tells me that, during the eruption, she heard echoes of the voices and recalled the experience of all the mentors and leaders that came before her. That lineage of scientific understanding could be traced all the way back to Jaggar, who more than a century ago decided his life's work should be dedicated to understanding volcanoes and protecting people who live in their domains from their occasionally lethal nature. "I think we did Thomas Jaggar and his vision proud," she says. But, she adds, it didn't stop her and her staff looking back at what transpired and thinking about what they could have done better. "That's part of the process. You look back, you do retrospectives, and you say: next time, we're going to do this differently."

IN NOVEMBER 2020, Kīlauea began to noticeably tremble. The volcano started to warp a little, and a spike of quakes at the summit suggested magma was on its way up. And finally, on December 20 at 9:36 p.m. local time, long before volcanologists suspected it would reappear, lava began cascading out of the walls of Halemaʻumaʻu, creating a lava lake more than 500 feet deep in just over two days.

Kīlauea was done dozing. Madame Pele was back.

II

THE SUPERVOLCANO

If volcanoes generally have a terrible reputation, then Yellowstone has a *really* terrible one. It is colloquially known as a *supervolcano*, a word that suggests it is capable of profound acts of volcanic violence. Scientists know that it is not about to bring humanity to its knees. But largely thanks to the unrelentingly enthusiastic screaming of tabloid newspapers and social media crystal-ball mystics, who frequently claim that Yellowstone is on the verge of a cataclysmic eruption, plenty of people treat it as if it were Earth's self-destruct button. Even during the height of the coronavirus pandemic, the *Guardian* reported that thousands were seen gathering around Yellowstone's famous geysers, not wearing masks and not physically distancing.[1] But I would bet that a fair few of them were asking National Park Service staff about their preparations for the supervolcanic end of days.

There is no denying that Yellowstone had a phenomenally destructive past. Its origin story is a saga of geologic annihilations: it comes from a long line of volcanic ancestors whose explosion craters were blasted into existence by an infernal jet that emerged near the planet's heart. This trail of scars is scratched across the western United States, stretching over hundreds of miles and back in time millions of years. But this volcano doesn't deserve to be seen as a harbinger of doom. That infernal jet won't persist into eternity. Nothing, not even supervolcanoes, last forever, and Yellowstone's halcyon days of eruptive fury may be long gone.

Everything you think you know about Yellowstone is probably wrong. It's not a killer-in-waiting. Supervolcanoes like it won't bring an end to civilization. If anything, Yellowstone is a lesson in how everything, even Earth's mightiest magmatic mountains and colossal cauldrons, are temporary on this spectacular, ever-changing world of ours.

Archaeological evidence suggests that people have been in the Yellowstone Plateau for 11,000 years or so. Native Americans have been familiar with the steaming geysers, corrupted soil, and broiling waters of the region for generations. But it wasn't until 1871, when Congress funded a game-changing expedition led by geologist Ferdinand Hayden to document the aesthetic delights of Yellowstone—through words, art, science, and photography[2]—that lawmakers and President Ulysses S. Grant were convinced to do something unprecedented: legislate to make Yellowstone the world's first national park.

Signed into law on March 1, 1872, the act[3] states that the secretary of the interior has exclusive control over the land and must ensure that the "timber, mineral deposits, natural curiosities, or wonders within said park" are protected for the purpose of "retention in their natural condition." Yellowstone National Park, all 3,472 square miles of it, existed even before the states it resides in—mostly Wyoming, with slim segments in Idaho to the west and Montana to the north—had joined the Union. Its early days were chaotic. The site frequently became a battleground as various Native American tribes tried to defend themselves against the bellicose US Army.[4] (Until 1916, there was no National Park Service.)

Geologically speaking, the park is an embarrassment of riches. Jefferson Hungerford, the brilliantly named park geologist, tells me that some of the rocks in the park date back to the Archean, a grand eon of time close to the beginning of the world. New rocks are still being made as minerals crystallize out of the superheated broths that gush from the park's 10,000 or so thermal features, including its many hot springs.

Grand Prismatic Spring may as well be the poster child for the entire park. Larger than a soccer field, it takes on the appearance of an interdimensional oasis. An ethereal, deep-blue hue paints its pupil, while adventurous bacteria thriving in the piquant waters fleck its iris in shades of yellow, orange, and green. In the winter, the snow sizzles into steam as it drifts onto the spring's rocky eyelid.

Other thermal features are more unwelcoming but no less remarkable. Sulfur Hills, on the north side of Yellowstone Lake, shows up as a bright red spot when viewed through a thermal lens. It has long been a land of fire and brimstone: eggy hydrogen sulfide gases effuse from the earth, an acid-stained dystopia hostile to vegetation and animals. "If you poke your finger through the crust, you'll get burned," says Greg Vaughan, a research scientist with the US Geological Survey. "You don't want to sit down and have a picnic there. It'll burn holes in your pants."

Yellowstone is also home to 500 or so geysers, half of all the geysers on the planet. One of them, Steamboat Geyser, shoots fountains of water up to 400 feet in the air. In the past, it has been nothing if not erratic, with pauses between major outbursts lasting anywhere from four days to 50 years. March 2018 marked a change: it began erupting as frequently as once per week, ultimately clocking up a record-setting 32 eruptions for the year. In 2019, it managed 48 eruptions, a tie with 2020.

It isn't clear what may have changed in the very upper slice of the crust, the rocky frosting on Earth's layer cake, to induce this uptick. Perhaps minerals precipitating out of the watery broth below the surface blocked the pathway to the surface, turning the shallow pool of fluids into a pressure cooker. The fluids, able to boil more quickly, could rocket up to the surface more frequently. Or perhaps the past few years of heavy snow provided the subsurface with an abundance of fluids in which to fuel Steamboat's plentiful convulsions. With no instruments tracking the dimensions of this complex cycle of fluid chemistry beneath the surface, definitive answers simply don't exist. But Steamboat's temperamental behavior is typical of almost all geysers, in that their plumbing is always being rearranged. And change is Yellowstone's hallmark.

Steamboat sits in Norris Geyser Basin. Around 115,000 years young, Yellowstone's oldest thermal basin is stupendously hot, reaching 460 degrees Fahrenheit just below the surface. It's as if a monster jabbed holes in the devil's kettle. The basin is a bazaar of hot springs, geysers, and fumaroles, hundreds of individual thermal features that heave, twitch, erupt, and effervesce. But these are sideshows compared to Norris's main attraction. Over the past two decades or so, an area larger than Chicago more or less centered on the basin has been sporadically rising and sinking by several inches, as if a sleeping beast underground was breathing in and out. Other parts of the park also inflate and deflate on similar scales, but at one point, the ground at Norris moved up at a rate of nearly six inches per year, the fastest ascent ever recorded at Yellowstone.[5] This is probably happening because hot circulating fluids keep getting stuck in the crust, causing it to rise, before finding a new escape path, allowing the crust to sink.

At Yellowstone, routine and repetition are anathema to the park. Entire thermal areas die off while others are born with remarkable speed. Just ask Vaughan, who has kept watch over Yellowstone for 12 years. Using a multitude of mechanical eyes in orbit around Earth, he has seen the park transmogrify, refusing to become static.

Back in 2017, he caught sight of a warm patch of dead trees in the backcountry, wooden corpses nestling in pale-yellow soil. Archived satellite and overflight imagery tracked their demise back to the late-1990s: their fate was sealed when a new thermal area cropped up and grew over the decades to the size of four soccer fields.[6] "A really large thermal area has been sneaking up on us for the last 20 years, and I just found it a couple of years ago," Vaughan tells me. Park rangers didn't spot it. It was only visible thanks to a time-traveling jaunt recorded from space. "That is a testament to how big and spread out Yellowstone is."

CHANGE, AS THEY SAY, comes from within. Lest we forget, Yellowstone is a volcano. All of its surface shenanigans only exist because two giant reservoirs of magma larger than entire cities remain in constant turmoil in the darkness.

At this point, we need a crash course in plate tectonics.

The planet's surface is a giant jigsaw puzzle. The individual pieces are tremendous slabs of rock the size of countries, continents, seas, or oceans. They are made of two layers. At the top is Earth's rocky crust, which comes in two flavors: the rocks that make up continents, and the rocks that form the seafloor. Below the crust is the mantle, a superheated, squidgy but predominantly solid 1,800-mile-thick layer. The tiny top slice of it, the rigid uppermost part of the mantle, makes up the base of tectonic plates.

These sandwiches of crust and upper mantle are named lithospheric slabs, but you may know them by the term tectonic plates. They sit atop the even squishier, more plastic-like mantle layer named the asthenosphere. Stiff tectonic plates drift on this more malleable lower layer of the mantle, and they all drift in different directions.

Tectonic plates have names. (Not names like Yasmin or Dave, sadly.) Much of America sits on the eponymous North American plate. This plate, like all other plates, is being jostled and flexed at its edges by the other plates surrounding it, from the Pacific plate to the west to the Eurasian plate way out east in the Atlantic. This means most of America's earthquakes happen along those boundaries, but there are exceptions to this. Western Wyoming, for example, is being very slowly pulled apart. This pulling creates earthquakes, many of which are imperceptible to humans.

That's great news for anyone wanting to give Wyoming a CT scan. Seismic waves produced by earthquakes travel at different speeds and trajectories through different materials. They zip through more rigid materials, like solids, but crawl through liquids. Hotter and slightly melted materials, less dense and less rigid than their colder counterparts, have lower seismic-wave speed limits. Scientists, being a savvy bunch, realized that this meant the journey of seismic waves, from quake to seismometer, could be used to work out what the underworld was made of: molten rock or frozen rock. Over the past few decades, scientists intercepted the seismic shouting being broadcast by countless temblors, and they assembled a view of Yellowstone's magmatic foundations as they went.[7,8]

Yellowstone has two connected magma reservoirs. The shallowest is 56 miles long and stretches from 3 miles beneath the surface to a depth of 11 miles. This elongated mass of rhyolite, a very gloopy type of magma, is more than four times the length of Manhattan.

Colossal though that may seem, it is the baby sibling of the two. Veins of molten rock stream down from the base of this upper magma reservoir, like strands of soft caramel, and connect with an utter monster: a more oblate mass of basalt—the runnier, hotter magma type familiar to Kīlauea—extending from 12 miles beneath the surface to a depth of 31 miles. It is four-and-a-half times bigger than the magma reservoir sitting above it, enough to swallow up the entirety of New York City.

Together, these reservoirs are the dyadic chambers of the engine that drives the dance of geysers at the surface. It isn't a direct link. The magma is not cooking the circling liquids and gases in the shallow subsurface. Instead, the heat from the upper magma reservoir seeps into its rocky roof, where it makes its way toward the surface. This residual heat bakes the rocks just shy of the surface, which provides pockets of water with the energy to boil.

I'm betting that a certain image has suddenly appeared in your mind, stolen from vague memories of high school textbooks or CGI animations on television: a huge cavern, packed to the rafters with a soup of molten rock, adorned with a big open tube reaching the volcano at the surface. The term used for these phenomena, *magma chambers*, itself reinforces the notion of a hollow lithological cathedral that accepts supplies of molten rock coming from deep below before emptying it from the top of its tallest spire.

But here's the thing: magma chambers don't exist. They have never been observed or detected by scientists.

If you leave a block of ice out in the Sun, it will quickly thaw and melt into a puddle of water. Rocks don't behave in the same way when exposed to sources of extreme heat. I would know, because I've thrown plenty of volcanic rocks into metal ovens designed to melt them. They don't become puddles. Instead, small bits of a volcanic rock will melt away at lower temperatures, while other chunks hold out until the mercury reaches incredible heights. That's because volcanic rocks are an assemblage of minerals, with each mineral having its own breaking—or, melting—point.

The same applies to Yellowstone's magma reservoirs. The deeper you dive into the crust, the hotter things get. But that heat isn't extreme enough at present to keep both reservoirs completely molten. Get a fresh injection of hot molten rock from below, say, and sure, more of the reservoir cooks and melts. But what you have for the most part is solid rock, with a network of molten ponds, pools, and slivers. Sometimes more of it is molten, sometimes less.

Think of it as a strange sponge, with the holes filled with a hellish gelatin. That gelatin, being hotter than the cooler, frozen volcanic crystals around it, is naturally buoyant, and the laws of physics demand that it rise as best as it can. It exploits zones of structural weakness, like big holes in the sponge, as it makes its ascent. Magma chambers aren't really cavernous gaps in the crust at all, but reservoirs of mush, serpents of partially molten rock confined within a labyrinth of crystals. They are Beelzebub's sponges.

Yellowstone's sponges may be gigantic, but they are mostly frozen right now. Seismic data suggest that the upper reservoir is between 5 and 15 percent molten. The lower reservoir is only 2 percent molten. Their dimensions are unquestionably mighty, but right now they are dormant, awaiting a return to the extraordinary eruptive activity that paints Yellowstone's past.

FIFTY-SIX MILLION YEARS AGO, the Farallon tectonic plate sat underwater far west of North America. Below it sat a fountain of superheated material,[9] emerging from the lower reaches of the mantle hundreds of miles down and rooted in place. As it rose, this fountain met the Farallon plate and melted bits of it. This fountain of mantle material, ascending into a lower-pressure realm, decompressed. The mantle melts when it decompresses, so this fountain's epic elevation triggered enormous amounts of melting on the Farallon plate's underside. Lots of magma was made. Lava erupted onto the ancient seafloor, creating cascades of sediment and volcanic rock over a dozen miles thick. This marked the fountain's opening salvo.

Fountains like this are named *mantle plumes*, and they can be found all over the planet. But, like magma reservoirs, mantle plumes have never been seen. Their presence has only been inferred from seismic waves bouncing off and bending through suspicious shadows in the mantle and from the behavior of the volcanoes above them. A protracted debate still rages over their properties, origins, and even their very existence. But most suspect that mantle plumes are real, epic, and capable of producing some extremely strange volcanoes. And many scientists suspect that an ancient mantle plume, once west of North America, began an extraordinary journey eastward 56 million years ago while also standing completely still; that is, the fountain was rooted in place and the North American plate slid over it on a roughly southwesterly path.

The Farallon plate, on a collision course with the North American plate, was doomed to be destroyed. As the two of them met 42 million years ago, the denser Farallon plate sunk into a trench in a process known as subduction. As the trench greedily gobbled up the Farallon plate, and both plates continued to move about above, the superheated fountain remained still rooted to the same spot in the mantle. It was if a lit match was held still under a piece of paper while the paper itself kept shifting about above it.

Around 17 million years ago, the Farallon plate—North America's tectonic shield—shattered under the relentless assault of the plume. In at least two places, the plume punched through the fragmenting shield. Reaching new heights, the plume could now cook the underbelly of North America. Rock melted with reckless abandon. And an infernal flood was unleashed.

Rivers of runny basalt lava gushed out of fissures opening up in the earth all over Oregon and Washington, as well as parts of Idaho and Nevada, some of it stampeding all the way into the Pacific Ocean. Fountains of molten rock and lava smothered the land, sterilizing everything in its wake. Nothing could have stopped it. As the lava began to cool, it entombed the Pacific Northwest beneath it.

In perhaps as quick as 800,000 years,[10] a geologic eyeblink, 95 percent of the Columbia River Flood Basalts had erupted. Lava had covered almost 64,000 square miles of land, some of it up to two miles thick. That's difficult to visualize, so think of it this way: all that lava could smother the entirety of Florida. If you packed all that lava into a cube, it would be 37 miles high, six times the altitude of passenger planes in flight.

At the same time, about 16 million years ago on the border of Nevada and Oregon, an explosion tore a hole in the ground 25 miles across as 240 cubic miles of ash, glass, and freshly baked volcanic rock showered the land,[11,12] As the North American plate kept moving across the face of the world, the stationary mantle plume below encountered unadulterated crust above it. More magma was made, creating vast underground bombs of gassy, pressurized magma. Bang: a million years or so later, another volcanic maw opened up on the triple border of Nevada, Oregon, and Idaho. Boom: three million years more, and a part of southwestern Idaho gets hole-punched.

And on it went. The North American plate, continuing to drift west-southwest by an inch or so each and every year, remained a moving target for the superheated fountain below. It kept blasting holes in the surface, leaving a track of scars up the eastern Snake River Plain. The largest eruption along this path may have been the recently identified Grey's Landing cataclysm on the Idaho–Nevada border, which threw 672 cubic miles of volcanic debris across North America 9 million years ago.[13] Squashed into a single cube, this pile of volcanic paraphernalia would be higher than Everest.

As the plume exploded its way into western Wyoming around 3 million years ago, a parabola of mountains surrounding the center of this hot spot warped and quavered, reacting to the fearsome fountain of fire between them. Today, the seismic shivering of that parabola can still be detected: the hot spot, the contemporary expression of the plume at the surface, continues to migrate to the northeast. Like a ship sailing through the sea, this is creating a bow wave of earthquakes in the "waters"[14] along and around its leading edge.

The hot spot arrived in what we now call Yellowstone 2.1 million years ago, long before humans arrived on the shores of North America. And that's when the sky vanished.

IT WAS THE 1950S, and as the world was once again changing beyond recognition, a Harvard doctoral student named Francis "Joe" Boyd decided to peruse the volcanic rocks of Yellowstone. He trekked through the park, taking a closer look at the volcanic rocks making up the walls of valleys, mapping outcrops of different geologies that appeared in a kaleidoscope of colors. Every now and then, he found what many of us would refer to as "dragonglass," but which scientists know as obsidian: a jet-black volcanic glass that breaks apart to form cutthroat-sharp shards. His mapping report, published in 1961,[15] notes that he found layers of obsidian up to 15 feet thick—easily enough dragonglass to defend yourself against an entire horde of White Walkers.

Although volcanic rocks were long known to be a feature of the park, work by others in the 1930s found that many of them didn't look like simple frozen lavas. Many of them seemed to have been dumped not as syrupy flows, but as downpours. They came from hailstorms so scalding that, even after they had been flung from a nearby volcano and had flown through the air, were still hot enough for the chunks of debris to weld themselves to each other, like bits of mozzarella on a pizza melting together to form one cheesy mass.

What Boyd had found was the lithified corpses of what a French mineralogist, François Antoine Alfred Lacroix, had seen at the turn of the century. Like Jaggar, he was struck by the May 1902 eruption of Mount Pelée. Lacroix watched as Martinique's volcano sent clouds of fury tumbling out of the volcano later that year. The clouds themselves were sporadically incandescent, so Lacroix referred to them as *nuée ardentes*, or "glowing clouds"—a term translated from an older Portuguese expression used to describe similar eruptions in the Azores.[16]

Today, these volcanic avalanches are known as pyroclastic flows, the Greek *pyroclastic* roughly meaning "broken pieces of fire." They are conjured in a few different ways. When a tall column of ash jets out of a vent, it ultimately loses its momentum when the explosive energy propelling it dwindles. Alternatively, the buoyancy of the plume, provided by its low density and the radiation of heat from its glowing volcanic matter, will fade over time as the plume cools down. Either way, what goes up must come down: the plume eventually falls back to Earth, generating pyroclastic flows. They can also appear when an eruption spills hot debris over its rim at any point during an explosive eruption or when a dome of thick lava being extruded from a vent becomes lopsided and collapses.

No matter their origins, the end result always inspires fear and awe in equal measure. They often rush down a volcano's slopes at speeds above 50 miles per hour, gliding on a friction-reducing cushion of air, a bit like pucks on an air hockey table.[17] They are a maelstrom of noxious volcanic gases and pieces of debris, ranging from microscopic particles of ash to clumps the size of boulders. Temperatures inside this torrent can hit 1,300 degrees Fahrenheit. They cannot be outrun, nor can they be survived. Sometimes, if they have sufficient oomph, they can flow uphill, so not even the high ground is safe in front of the flanks of certain volcanoes.

In the 1950s, the science of pyroclastic flows was in its infancy. Today, our knowledge is still in its adolescence, partly because their unpredictable appearances and brute force make them difficult to study. Boyd's naturally soldered-together volcanic deposits were baffling, with various scientists suggesting several ways in which they could form. Pyroclastic flows were a strong candidate. Boyd himself concluded that these flows made much of the welded stuff he found all over Yellowstone. Not only were they extremely hot when they emerged, but also these pyroclastic flows must have still been over 1,100 degrees Fahrenheit even after they stopped barreling across Yellowstone.

Boyd also thought they erupted from a number of different vents. He didn't find these vents, but he did casually note that part of the Yellowstone Plateau appeared to be a depression of some kind, as if the ground fell in on itself long ago. He didn't link the two, but having painstakingly mapped out the chilled remnants of an ancient dragon's breath, he had unknowingly stumbled upon one of the volcanic wyvern's former lairs.

From 1966 to 1971, a US Geological Survey group went all-in on Yellowstone, spending warm summer months updating geologic maps of the area and looking at minuscule rock fragments beneath microscopes to untangle their chemistry. NASA partly funded the endeavor. At the time, NASA was developing instruments for satellites that could carefully study various aspects of the planet from space. The agency thought detailed maps of Yellowstone would be good to compare to images snapped by its own orbiting robots.[18]

The first report on this effort, published in 1972,[19] not only confirmed Boyd's suspicions but also found three dragons' lairs within the park: a trio of rocky cauldrons, each the site of a furious eruption, and each responsible for creating its own pyroclastic flow bonanzas. Half a century later, scientists are still trying to piece together the story of those three eruptions.[20]

As ever, uncertainties abound. What is indisputable, though, is that the volume of volcanic debris produced by this eruptive trilogy is enough to make anyone's jaw drop: all three cataclysms combined would easily fill up the Grand Canyon.

Around 640,000 years ago, one eruption depleted a magma cache so massive that 240 cubic miles of material was jettisoned across the realm. About 1.3 million years ago, another outburst threw 67 cubic miles of volcanic flotsam and jetsam into the air. But the first in the Yellowstone trilogy remains triumphant. The mantle plume had finally allowed enough magma to accrue beneath Wyoming to put on a show-stopping performance—and 2.1 million years ago, Yellowstone supervolcano was born when 588 cubic miles of shattered fire slammed into North America, a volume almost 10,000 times greater than all the matter unleashed by the infamous May 18, 1980, eruption of Mount St. Helens in Washington State.

No one was there the day the first eruption began. But thanks to more than 50 years of research, observations of similarly sized eruptions in recent times, and a sprinkling of reasonable assumptions, we know the eruption sequence was gargantuan. I say "sequence" because eruptions are rarely single events. They are usually extended features. There were multiple stages to the 2.1-million-year-old eruption: distinct moments of chaos between moratoriums where considerable volumes of ash snowed out of the sky and pyroclastic flows thundered across the land, whose collective remains are known as the Huckleberry Ridge Tuff.

Scientists often treat the eruption as one single event because it happened such a long time ago. But scientists such as Colin Wilson and Madison Myers, volcanologists at Victoria University of Wellington in New Zealand and Montana State University, respectively, have begun to identify minuscule gaps in time preserved within the Huckleberry Ridge Tuff. Along with their colleagues, they have spent years examining the rubble left behind by Yellowstone's first major volcanic salvo through a bevy of high-tech magnifying glasses. They have gleaned the tiniest volcanic remnants to tease out their chemical ingredients and find out which specific batch of magma brewed them up.[21] Small structures and distinct chemical signatures found between layers of volcanic rock betray the existence of pauses[22] between each phase of frenzied volcanism, when ash was being washed away by streams or covered by rainwater. These pauses vary in length from hours and days to years and decades, meaning the Huckleberry Ridge Tuff wasn't formed during one very action-packed week. Instead, the duration of Yellowstone's first supereruption could have been over the length of an entire human life span.

"How you quantify that time in a way that people trust is very challenging," says Myers. But quibbles over hours and months aside, the fact remains that this supereruption was segmented, not singular. "And what's interesting to me," she says, "is to try and envision what it looked like."

EVERY NOW AND THEN, the earth would jolt and jerk. Over weeks, or even months, the drumbeat would quicken and get progressively stronger. Cracks would snake through the ground.

Deep below, a magmatic infusion was pooling closer to the surface than it had been for millennia. Individual lenses of melt coalesced and headed upward. At such shallow depths, gases previously dissolved in the solution began to escape, turning into bubbles and adding pressure. Parts of Yellowstone would noticeably deform. The ground would rise in one area while another would recoil. Toxic gases and steam would whoosh out of fumaroles at remarkable volumes. Rocks, faced with forceful magma, would shatter.

Something had pulled on a geologic trigger. Perhaps a magmatic missile, breaking through to the surface, had too much momentum to be stopped. Maybe a major earthquake, caused by the stretching of Wyoming, created a spider's web of underground fractures that liberated the magma from its trap. Either way, one day, a cupola of magma found a route to the surface world. A vent opened up in the earth, and liquid fire was met by the cool air. The once-trapped gases within the magma explosively decompressed, throwing millions of flecks of lava into the air, quickly quenching into microscopic particles of glass. Columns of ash, sustained by their inner fires, kept them climbing into the sky. For hours or days, ash fell on Yellowstone as the ground rumbled and the air sizzled. Rain may have washed some of the blanket of ash away, turning it into a mobile slurry as thick as concrete.

Then, a second cupola of magma exploited its own path to freedom. Another explosion, somewhere else in Yellowstone, cast everything aside as a second plume of ash polluted the skies and showered out of the land. Perhaps days later, the pin of a third cupola-shaped grenade was pulled. Ash greeted the sky for a third time as animals scarpered. Multiple ash cannons created a fountain of volcanic matter. Ash from one vent sometimes fell alone. Other times, it was accompanied by others, like a mistimed fireworks display.

And this was only the beginning.

The transition from ashy snow into full-blown cataclysm would now be under way. The ascent of an enormous vessel of molten rock was unstoppable. The broken rocky roof trying to hold it back would fail under the strain and heat. Unchained, unleashed, the magma gathered right at the top of the crust. The ground above ballooned. Cliffs began to crumble and turn into landslides. The flow of rivers would have been interrupted, perhaps even reversed. Boiling water would be bursting up through seams in the rock. Earthquakes would be tearing down trees, frightening animals, sending ripples through the air.

It may have begun at night or in the middle of a Sun-drenched afternoon. Somewhere around the northwest corner of what is now Wyoming, the earth, and anything attached to it, would have been annihilated by an explosion or series of explosions so powerful that everything—creatures, plants, rocks—would have been vaporized. The energy of the eruption would easily dwarf the sum total energy of every single explosive used during both World Wars, including both atom bombs. Volcanic rocks, flowing like liquid plastic, would have been shot outward at speeds approaching supersonic.

Around the edges of the explosion, broken volcanic fires would be boiling over the rims of the patchwork earth. The dragon emerged from its lair, and exhaled: pyroclastic flows the size of towns gushed forth, barreling down into valleys and soaring over hills and up mountain ridges. Their bulkier lower layers sheared the land like a scythe, while their upper layers left roads of ashy glass in their wake. Anything living caught in their bulldozing would have been instantly embraced by death: the water in skin and muscle would immediately evacuate as trillions of cells ignited; skeletons would boil in their biological casings; organs would enter a state of shock, then shut down, all within a matter of seconds. Entire forests would have been decapitated, pulverized, and enflamed all at once. These glowing clouds, stopping only when sapped of sufficient momentum, traveled dozens of miles from their source.

At the heart of this volcanic onslaught, a sustained and colossal column of fine volcanic debris rose into the sky, punching through any clouds that stood in its way. Viewed from space, a terrifying umbrella of black and gray would be seen spreading out more than a dozen miles above the ground. If it were daytime, the light of the Sun would be consumed. If it were night, the glowing moon, should it dare to come out, would be drowned in darkness.

As charged particles of ash bounced off each other, driven apart and occasionally ripped into pieces, a huge electrical imbalance accumulated in the column of ash. Giant bolts of lightning began flashing through the storm, tearing the air into confetti and broadcasting bone-shaking booms of thunder out to the wider world. As the days went by, some of this ash would start circling the globe, stealing away the glimmer of stars, blocking out the rays of our own, and rendering parts of the planet temporarily cold. Corrupted by widespread volcanic gases, weather systems in the Northern Hemisphere would be briefly short-circuited; rain would fall out of season, and impromptu droughts would dessicate normally saturated corners of the globe.

With the stockpile of buried magma severely depleted, the ground's foundations would have evaporated. Not long after the first pyroclastic flows emerged, the ground fell in on itself, tumbling into a serrated cavity hundreds of feet deep, eating up anything in the region that had defied the odds and remained standing.

After the glassing of the landscape was complete, the eruption was drained of its deleterious propulsion. Rain, bouncing off the ceilings of pyroclastic flows—now drained of their fire and forward momentum—hissed, the water boiling off as steam.

The ground had been torn open. The lair was empty. The dragon had vanished.

Weeks, perhaps months, passed. The Sun returned. The land, once abundant with life, was now the aftermath of a great conflagration. Death lingered in the air.

Then, preceded by rumblings not so deep below the surface, the volcano began its encore. Surges of volcanic debris smothered the earth once again, earthquakes, fire, and all—and ash climbed into the sky for a second time. As before, the roof of the expired magma tank collapsed in on itself, sculpting out a second sunken cavity in the land.

Years went by, perhaps decades. The ash was washed away. Animals venturing into the area would have no recollection of these two tremendous acts of destruction, ones memorialized by pyroclastic flows, now cold, turned into curious monuments of stone. Flowering plants sprouted. Saplings began to grow into trees. Animals returned. Life had begun to sculpt this brave new world into a home once again.

And, for a third time, this pioneering life experienced the same doom as its predecessors. It would have been exterminated by armies of pyroclastic flows, devoured by liquid fire and scorched air. Ash clawed at the sky before hailing back down, polluting waterways and soil, killing off plant life hiding beyond the horizon. Shards of volcanic glass, drifting on the wind, would have found its way into the overworked lungs of animals near to the blast, clogging up their respiratory apparatus and killing them from the inside out. And once again, when all was said and done, a deep and empty cavity was left behind.

This 2.1-million-year-old catastrophe wasn't just a single explosion created by a solitary subterranean volcanic wyvern. It was a tripartite eruption driven by multiple magmatic lakes. And its wreckage can still be found, lingering across America, millions of years later.

Not long after the last collapse had happened, its pyroclastic flows froze into cliffs of welded rock. All three invasions of broken fire had been stacked, one atop the other, forming a combined layer more than a thousand feet thick in places and covering 6,000 square miles of land. The ash made throughout the entire eruption sequence was woven into a veil and draped over the entirety of continental America: scientists have found ancient black snow all over, from the Canadian border to Mexico, from the waters of the Californian coastline to the middle of Texas. Those three separate geologic puncture wounds may have happened in the same place, causing the pit to deepen and enlarge with each successive blast. They may have instead been near enough to each other than they ended up overlapping. Either way, the total sum of their destruction was a distorted, circular-ish chasm in Yellowstone 60 miles across.

The next major eruption at Yellowstone, 1.3 million years ago, would be a small fry compared to the original. The one after that, 640,000 years ago, would be a return to violent form. Between these three big blasts, and ever since its third show of force, Yellowstone has seen countless flows of stodgy lava emerge from the ground. The last major magmatic outpouring took place 70,000 years ago, its largest flow crafting the Pitchstone Plateau in the southwestern section of the park. But nothing would ever come close to matching that elongated moment in time 2.1 million years ago, when a schism in the continent burned the earth and blackened the sky, leaving behind nothing but ghosts of ash.

This saga can only be told thanks to the tireless work of a multidisciplinary coalition of scientists. "It's kind of beautiful how much detail you can get out of it and the story you can tell," Madison Myers says. Historical explosive eruptions, like the 1980 Mount St. Helens blowout, are themselves jaw-dropping: true "holy shit" moments, she says. But Yellowstone's first Mephistophelian paroxysm makes even the most vigorous offerings from contemporary volcanoes seem relatively tame.

MIKE POLAND, TO MY EARS, is never anything less than cheery. We've spoken about Yellowstone plenty over the past few years, and he's always game to chat. Volcanologists, by and large, are usually happy to pick up the phone and speak to me, but conversations with Mike are delightful, breezy, animated, informal. His job is anything but. Since September 2017, he has been the scientist-in-charge at the Yellowstone Volcano Observatory, a facility of the US Geological Survey. Although he spends much of his time monitoring and studying the geologic idiosyncrasies of the eponymous volcano, a disproportionate amount of his day-to-day is spent fending off the notion that the magma reservoirs under northwestern Wyoming are a geologic weapon of mass destruction impatiently waiting to detonate.

I always experience a soupçon of guilt whenever I add to the mountain of madness by dropping the word *supervolcano* into our conversations. He hates it like millions hate the soapy flavor of coriander, that much is clear. The word itself has a few origin stories, but according to Poland,[23] a review of a 1948 book on Oregonian volcanoes used the phrase. Scientists didn't adopt it, and the media didn't run with it. But in 2005, a Canadian–British docudrama called *Supervolcano* was released, showing what may happen if another massive eruption rocked Yellowstone. Things looked grim and frightening. The team behind it consulted with, among others, scientists at the Yellowstone Volcano Observatory, so some segments of it, although speculative, are fairly accurate. Others, less so—but it mattered not. The splashy visual aesthetic of it all helped thrust the titular term into the public spotlight. Some scientists began using it in their peer-reviewed papers around that time too, because the word does have an undeniable ring to it.

But a supervolcano isn't what you think it is.

There exists a scale named the Volcanic Explosivity Index (VEI),[24] developed by volcanologists Chris Newhall and Stephen Self in 1982 in an attempt to quantify the magnitude of explosive eruptions, past and present. A "0" rank is assigned to eruptions that effuse a modicum of lava and debris and have ash plumes no more than a few hundred feet high. A "3" rank is several orders of magnitude more prolific than a "0": plumes reach several miles into the sky while the eruption chucks out a decent amount of debris, plenty to fill in the streets of a nearby town.

As you scale up, the number of these eruptions worldwide get ever rarer. A VEI 8 eruption, the very top of the scale, is anything that produces at least 240 cubic miles of volcanic material during its explosive eruption. (Why 240 cubic miles? It's pretty arbitrary, but 240 cubic miles equals 1,000 cubic kilometers, and scientists love round numbers.) A VEI 8 eruption is known as a supereruption. Any volcano that is known to have produced at least one VEI 8 in its geologic lifetime is granted the title of supervolcano. And that's all there is to it.

Yellowstone is a proper supervolcano. Its eruption 2.1 million years ago (588 cubic miles of volcanic matter) easily qualifies, as does its most recent serious flare-up 640,000 years ago (240 cubic miles), although only just. Its 1.3-milllion-year-old throwdown (67 cubic miles) doesn't cut it. That one's a paltry VEI 7.

However, Yellowstone is far from the only supervolcano on Earth. Toba, a volcano on the Indonesian island of Sumatra, chalked up a supereruption 74,000 years ago. The eruption of Taupo, a volcanic pit in the middle of New Zealand's North Island, turned itself into a supervolcano 26,500 years ago.

Each matched up to Yellowstone's own earth-shattering prowess. Taupo's—the world's most recent VEI 8—carved a 22-mile hole[25] in the ground as more than 270 cubic miles of magmatic shards flew across the archipelago.[26] Pyroclastic flows buried much of the sizable North Island in 660 feet[27] of hot chaotic debris, while nearly 400,000 square miles of land was dusted in a layer of ash.[28] The Chatham Islands, 600 miles from Taupo, were buried in a blanket of ash seven inches deep.[29] Toba's VEI 8—the largest volcanic eruption in the past 2.5 million years, perhaps longer—was a real monster, carving out a great deal of today's 62-mile chasm in Sumatra.[30] Around 672 cubic miles of volcanic material blasted into the atmosphere, while four inches of ash coated 2.7 million square miles of land. That's enough to shroud nearly nine-tenths of the lower 48 US states.

The immensity of these explosions beggars belief. In the summer of 1962, America buried a nuclear device eight times more powerful than the bomb dropped on Hiroshima below the arid sandy surface of the Nevada desert. It carved out the largest artificial crater in America, a bowl 1,280 feet across.[31] The caldera mostly excavated by Toba's 74,000-year-old kaboom is 256 times longer. It's like comparing a firecracker to the Holy Hand Grenade.

Some scientists consider a volcano to be geologically (not necessarily historically) "active" if it has erupted at least once during the past 12,000 years or so, the start of a period of time known as the Holocene. According to the Smithsonian Institution's Global Volcanism Program, the Library of Alexandria for the world's volcanoes, Taupo has erupted 25 times during this epoch, so it's considered active. Yellowstone's last major eruption was 70,000 years ago, but bits of evidence for a handful of convulsions in the current epoch just push it over the line.[32] Unlike the cataclysms in their distant past, none of these contemporary eruptions have come close to being VEI 8's. But remember: once a supervolcano, always a supervolcano. And therein lies the problem.

Supervolcano evokes images of a source of seething eruptive rage and brutality. But are supervolcanoes the real threats to humanity? Not even close. A supervolcano refers to a volcano—active, dormant, or extinct—that has erupted 240 cubic miles of debris *just once*. This doesn't mean it will produce another 240 cubic miles of material during future eruptions. Think of supervolcanoes as Olympic athletes: they may have won gold once or twice in the past, but it doesn't mean they will keep on winning gold. They might have retired.

Yellowstone has had two supereruptions. But for most of its lifetime, blasts caused by trapped steam and lava flows isolated to the park have been far more common manifestations of its volcanism. Right now, the majority of Yellowstone's magma reservoirs are solid. There is no indication that it is in any state to erupt in any manner whatsoever, let alone produce another supereruption.

Its glory days could very well be consigned to geologic history. The North American plate is continuing to drift southwest at the same rate your fingernails are growing. The volcanic hot spot will keep migrating northeast. In a couple of million years, when it heads under the Absaroka mountain range on the Montana–Wyoming border, it will sit under thick, ancient continental crust, a far cry from the thin, stretched-out crust it has been dealing with in Wyoming. "Will the hot spot be able to burn its way through?" says Poland. "We may have seen the last of Yellowstone's volcanism, more or less—you know, the really big, impressive stuff."

It may depend on the strength of the blowtorch. Victor Camp, a volcanologist at San Diego State University, is one of several scientists decoding the mysteries of said blowtorch. His work suggests that it is, in his words, "moderately strong." The volume of magmatism may diminish, but he speculates that the plume may be potent enough to cook Montana's rocky base and birth volcanoes at the surface.

This process of migratory volcanism, of waxing and waning volcanic activity powered by a mantle plume, isn't especially rare. Although the associated volcanism is quite different from that seen in the western United States, another mantle plume beneath the wandering Pacific plate is thought to be responsible for creating the volcanic Hawaiian archipelago, too. It has created a warped but continuous line of volcanoes over millions of years, with each older volcano dying out as the plate shifts and the plume gets to build a new volcano. The hot spot is currently centered roughly where Kīlauea is now. One day, Kīlauea will be too far from the plume that birthed it to erupt, and it will die out.

Nothing lasts forever, including volcanoes. Not even Yellowstone supervolcano is immune. You could say that it is dying. The odd lava flow or two in the park aside, its supereruptive history may never be repeated. Sorry, doomsday fans, but what you see today is what you may get: a glorious collection of effervescent pools, fizzling geysers, and tumbling waterfalls.

Every now and then, a swarm of earthquakes—a bunch of tremors happening near each other in a usually short space of time—occur in Yellowstone. The crust is always being stretched by tectonic forces, and hydrothermal fluids are always circulating below, cracking open bits of rock. Jamie Farrell, a seismologist at the University of Utah who listens to Yellowstone's heartbeat, says that swarms are "just everyday life for an active volcano." He registers anywhere between 1,500 and 2,500 quakes per year in the park, with 40 to 50 percent occurring as part of swarms. "They are a normal way for earthquakes to occur in volcanic areas. And not just these really large volcanoes. All volcanoes typically see these swarm activities," he explains. That sadly doesn't prevent some from gesticulating wildly and falsely framing this shaking as a prelude to volcanic fury.

This click-harvesting effort reached a perversely hilarious nadir in 2018 when a British tabloid ran an article[33] with the title "Yellowstone volcano ERUPTION warning: Hundreds of bison dead as fears of mega blast grow." I kid you not: the tabloid explicitly made a link between a bison cull in the park and the possibility of a major eruption. Did the bison know something we weren't meant to hear? Were they silenced to keep a conspiracy from emerging? I share the headline with Poland. He laughs for a good 20 seconds—loud enough that my dog, Lola, wakes up from her nap and quickly scampers in from the other room to investigate the ruckus.

The US Geological Survey isn't that bothered about another supereruption. In 2018, it published the latest version of its National Volcanic Threat Assessment.[34] In it, USGS scientists assessed the hazardous nature of each of America's 161 active volcanoes, taking into account a huge number of factors: how explosive and destructive have a volcano's eruptions been in the past; whether it is capable of producing pyroclastic flows or not; whether it is covered in water or ice, which can mingle with rising magma to produce unpredictable explosions and prominent ash plumes; whether the volcano is far from or near cities, airports, power plants, waterways, or other vital infrastructural sites; and so on. The volcanoes were then ranked and sorted into five broad categories, from "very high threat" to "very low threat."

Kīlauea, capable of producing prolonged, city-threatening lava flows, took the top spot. Washington State's Mount St. Helens, which erupted out of its side 40 years ago and melted plastic 13 miles away from the volcano as the air burned, came second. Yellowstone came

21st. An eruption of some kind in a populous national park surrounded by satellite towns would, of course, be awful. But does this ranking really suggest that the mere mention of this particular supervolcano should send our amygdalas into overdrive?

Both Kīlauea and Mount St. Helens, highly active and eruption-prone volcanoes, are far more likely to threaten people, both directly and indirectly. Both erupt in very different ways, but they erupt far more frequently and far closer to people and infrastructure than Yellowstone ever has or ever will. The 1980 eruption of Mount St. Helens may have killed 57 people and caused $2.7 billion in damage, but it was only a pathetic VEI 4 (or 5, depending on who you ask): a regular eruption. Taupo may have had a supereruption 26,500 years ago. But 1,500 years ago, it was rocked by an eruption 100 times bigger than the 1980 Mount St. Helens event. It covered the land in pyroclastic flows 330 feet thick, which vaporized everything in its path and completely destroyed forests 20 miles away while blanketing all of New Zealand in an ashy sleet. That eruption, clearly spectacular, only took silver: a VEI 7, not a supereruption. And who cares about second place?

OBVIOUSLY, THIS IS DAFT. If you were about to be consumed by any of these eruptions, you wouldn't care about whether they were a genuine supereruption or not. Janine Krippner, a volcanologist at the Global Volcanism Program, puts it best. "There is always an obsession with the biggest thing," she says. "But the biggest thing is usually the rarest." That's disappointing in many everyday situations, but a relief when it comes to Earth-shattering explosions.

Don't get me wrong. Yellowstone has its dangers. They just aren't what you expect. "The most dangerous thing you do in Yellowstone is drive in your car," says Farrell. Whether you crash into a ditch or skid off the road in the icy winter, the odds of you being injured or perishing in a vehicle accident is always far higher than you probably acknowledge. Deciding to wander off trails and into the backcountry may also bring you face-to-face with some of the park's wildlife, including its grizzly bears. Unpleasant encounters do happen: according to the National Park Service, there have been 44 grizzly-based injuries since 1979.[35] The service also reports that, since 1872, there have been eight bear-induced fatalities. That may strike you as worrisome. But during that same period, 121 people drowned after falling into one of the park's many lakes and streams, so being clumsy is a riskier trait to possess than a penchant for wild bear hugs.

Speaking of being clumsy, Yellowstone's hot springs provide a terrorizing trap for the uncoordinated and reckless. Many spit and fizz at near-boiling temperatures, and several are incredibly acidic. Sulphur Caldron, for example, approaches the pH of battery acid. Stepping off the boardwalks and falling into such a witches' brew initiates a terrifying act of extreme exfoliation: your skin is chewed away by the acid as the high temperature blows apart your blood vessels.

Several visitors to the park have found this out the hard way. In 2020, a woman illegally entered the park while it was still on lockdown, subsequently falling into a scorching pool near Old Faithful Geyser as she was walking backward to set up a photograph.[36] She survived but had to be helicoptered to a burn center in Idaho.

An incident in 2016 is pure nightmare fuel. A brother and sister were hiking through an off-limits part of the park, searching for a thermal feature cool enough for them to dip in. During their quest, the man slipped into a hot spring in Norris Geyser Basin and was unable to get out. Unimaginably painful high-temperature burns or his organs shutting down due to thermal shock quickly killed him. Park rangers went to recover the body, but inclement weather conditions and the volatility of the hot spring prevented them from doing so. When they came back the next day, his remains had disappeared. Lorant Veress, a Yellowstone deputy chief ranger, told reporters at the time that "in a very short order, there was a significant amount of dissolving." His wallet and flip-flops, however, survived the ordeal.[37]

You are less likely to be killed by one of the park's hydrothermal explosions. Every so often, the plumbing for a hot spring or geyser gets clogged up. Sometimes, subterranean fluids underground come up against an immovable cap of rock; they are above their boiling point, but they are squashed up so tightly that bubbles cannot form. A single crack in these lithic lids is like blowing a hole through a pressure cooker: the fluids explosively decompress, with liquid water flashing to steam and turning its surroundings into a demonic sauna faster than you can blink. Rocks are both flash-fried and shorn apart. As the December 2019 tragedy at New Zealand's Whakaari/White Island epitomized, these events are for the time being essentially unpredictable and, if people are in the way of the explosion, lethal.[38]

Yellowstone sees around one small hydrothermal blast per year, says Farrell, mainly in the backcountry. But deep, old scars demonstrate that, very rarely, the park experiences more significant steam detonations, akin to "an entire geyser basin exploding at once." Yellowstone is in fact home to the largest hydrothermal explosion crater in the world: a 1.5-mile coliseum on the north edge of Yellowstone Lake, which came about 13,800 years ago.[39] Another would prove fatal today to anyone in the blast radius, but the odds of this fate befalling anyone on your average day remains considerably low.

The park's many earthquakes are also unlikely to harm you—unless, of course, tectonic forces permit a bigger seismic sibling to visit. On August 17, 1959, three different faults jolted forward, and a magnitude 7.3 temblor shook Hebgen Lake, just west of Yellowstone. It killed 28 people, most of whom met their ends via a large landslide.[40] Scientists cannot say for sure, but a band of seismicity extending from Hebgen Lake to Norris Geyser Basin observed in the present may be aftershocks from that initial 7.3 mainshock half a century ago, like a bell still ringing long after it was struck.

That disaster was "kind of a wake-up call," says Farrell. Scientists slathered the park in seismometers. Originally run by the US Geological Survey but later taken over by the University of Utah in the mid-1980s, the network today consists of 30 cutting-edge seismometers that can detect the ground trembling in any direction. Along with the dozen or so other seismometers operated by other scientific agencies and a number of older seismometers still in place, Farrell says that Yellowstone has "one of the best seismic monitoring networks on any volcano." That's a relief, because geologically speaking, he says, "the seismic hazard is by far and away the largest." Big quakes are rare but have top-tier maiming potential.

Dangerous volcanism of any kind is right at the bottom of the likelihood list. An eruption—an outpouring of lava or even a moderately explosive event—could transpire in the far-flung future, depending on what the magma mush below does over the years. Fortunately, Yellowstone's volcanism is monitored to high heaven. The US Geological Survey has five volcanic observatories across the country: in Hawai'i, the Cascades (a spine of volcanic mountains running up the western United States), Alaska, California, and Yellowstone, each designed to monitor its own regional volcanic kingdom. Like the others, Yellowstone's is made up of partnerships. The US Geological Survey works hand-in-hand with, among others, the National Park Service, the National Science Foundation, the Universities of Utah and Wyoming, Montana State University, the state geological surveys of Idaho, Montana, and Wyoming, and UNAVCO, a consortium of universities that facilitate geoscientific research.

Very few volcanoes erupt without putting on a show first. A series of earthquakes, their seismic rhythm indicative of magma breaking through rocks, would be detected in advance. "It's pretty hard for magma to move through the crust without breaking something," says Farrell. The park's GPS stations would also track significant ground deformation should magma be pooling at shallow depths. And gas monitors would sniff out volcanic fumes as they escaped from the magma as it rose.

A huge number of scientists are on the case. If an eruption was on its way, they'd see it coming. They would be frantically waving their arms about and shouting from the rooftops, not conspiring to cover it all up. "Even if we were deleting earthquakes right and left, there are tens of thousands of people that live in and around the park," says Poland. "They'd, uh, notice."

WE ALL SECRETLY LOVE it when we watch a movie or read a book and the worst-case scenario happens. And I know you secretly want to see the worst-case scenario at Yellowstone play out. So let's do it.

In the unluckiest of timelines, involving the most improbable odds, Yellowstone could produce a third supereruption at some point far into the future. Certainly, anything close to matching its 2.1-million-year-old magnum opus would be an unprecedented disaster. Pyroclastic flows would swamp the park, immediately killing anyone there. But the real danger comes in the form of the ash falling out of the plume.

If a high-intensity eruption is churning out plenty of volcanic matter into the sky, it'll likely form an umbrella. During the Huckleberry Ridge Tuff–forming eruption, plumes of ash would have risen so high that the ash would have made it into the stratosphere: beginning at around seven miles above your head, this slice of atmosphere sits directly above the layer that contains the weather. And when the plume reached a height in which its density matched that of the rarified air into which it had invaded, it would spread out left and right, forming an umbrella shape.

Alexa Van Eaton, a volcanologist at the US Geological Survey's Cascades Volcano Observatory, says that a lower-intensity but nevertheless prolonged eruption could still jettison plenty of ash over a wide area if the local wind conditions permitted it. But, she adds, having an "umbrella cloud is a game-changer." The faster the volcanic matter is erupted, the faster the umbrella cloud spreads, allowing it to overpower the swirling winds and push out over a broader area. That means in the event of a VEI 8 at Yellowstone, much of America would be met with a blizzard of ash.

A 2014 study, co-authored by Van Eaton, used computer simulations to see what would happen if 80 cubic miles of ash was jettisoned out of Yellowstone over days, a week, or a month. A thin dusting would reach the East, West, and Gulf Coasts; several inches to a foot or so would fall in the northern Midwest states; several feet of ash would cover the northern Rocky Mountains region.

Claire Horwell, director of the International Volcanic Health Hazard Network, says there will be health problems: those with preexisting lung conditions may suffer from serious ailments if they inhale sufficient ash. But in general, ash fallout "is more of a psychological concern" than a physiological one. The sight of the dimmed Sun, and a snowstorm of gray glassy matter sifting down across the entire region, would strike dread into anyone's heart.

If Yellowstone's VEI 8 event is spread over weeks, months, or even years, people's sustained exposure to volcanic ash may cause health problems. It's difficult to say, though, because research on the long-term effects of volcanic ash inhalation is extremely limited. Pneumonoultramicroscopicsilicovolcanoconiosis—a word my spell-checker miraculously identified as legitimate—is the name of the lung disease people may get by inhaling crystalline silica dust, a by-product of certain eruptions. But the condition has never been officially diagnosed. It is hypothetical, and crystalline silica may not even be toxic. "The jury is just completely out on that," says Horwell.

The real damage would be to America's infrastructure. The ash from the 1980 Mount St. Helens eruption clogged waterways, short-circuited electrical wiring, blocked roads, and grounded flights. The amount of ash produced by a VEI 8 at Yellowstone would be incomparably more disruptive. Crop fields would be crushed under the weight of the ash, and fertile soil would be toxified. Car engines would clog up and fail. Overhead wiring and cell-phone towers would malfunction.

Roads and rail and air transport would be unable to function. Weaker roofs would collapse under the weight of several feet of dense ash. A series of studies[41,42] looking at the effects of an eruption in the city of Auckland in New Zealand underscores how every facet of your life would be perturbed. Sewage outlets, something we all take for granted, would clog up, requiring years—years!—to fix, a distinctly unhygienic problem you wouldn't wish on your nemesis. The economic damage would also be profound, likely causing grim socioeconomic ripple effects around the world.

But the world won't end. It would not even come close to bringing civilization crashing down. We know this, because this experiment has already been run.

THE 74,000-YEAR-OLD supereruption of Indonesia's Toba was a monstrosity. It produced a lot of sulfur dioxide, a common volcanic gas that, when mingling with stratospheric moisture and sunlight, can clump together to form a substance known as an aerosol—in this case one that is excellent at reflecting sunlight back into space. In large enough quantities, volcanic aerosols can cool the planet a little for a short period of time.

A 2009 paper[43] modeling the ash and gas dispersal from Toba's supereruption suggested its aerosols could have caused a global temperature drop of 14 degrees Fahrenheit,[44] a chill that would have taken decades to recover from. At the time of publication, genetic evidence implied that only a limited population of modern humans were leaving Africa 60,000 years ago. Together, both lines of evidence were thought, by some, to suggest the global chill had disrupted our ancestors' environments so detrimentally that it almost wiped them out. But others were skeptical.

In 2003, archaeologists were following up on the idea that modern humans, many millennia ago, had first migrated into India in the Jurreru River valley of southern India. While there, they were told about villagers mining thick deposits of volcanic ash in the valley, selling it off for various industrial purposes. When they took a closer look, they found ash with the geochemical signature of the supereruption of Toba, a volcano thousands of miles away.

Michael Petraglia, now a researcher of human evolution and prehistory at the Max Planck Institute for the Science of Human History in Germany, was a member of the team. Despite its distance from the cataclysm, he says, such deposits show that Toba's ash would have blanketed all of the Indian subcontinent in a layer two inches thick. But that wasn't what blew Petraglia away: identical archaeological remains were found above and below the Toba ash layer, showing that these people, whoever they were, survived the supereruption. "That was stunning," he says. "That was an amazing moment."

Comparing the buried artifacts and tools to others found around the world from that time period, it became clear they were closely related to the Middle Stone Age of Africa, a time defined by *Homo sapiens*. In other words, those survivors were us.[45]

In the subsequent years, the severity and longevity of this thermal nosedive was repeatedly revised down.[46] Ash and organic matter salvaged from Lake Malawi in East Africa showed no evidence of a volcanic winter at the time of Toba's roar.[47] And in 2020, Petraglia and his colleagues, working in northern India, found much the same evidence for continued human existence[48] as they did back in southern India at the turn of the century.

Today, few buy the near-extinction hypothesis. Toba's supereruption was no doubt breathtaking, a true display of Earth's volcanic wrath. But it was not enough to smite our primitive predecessors. "We did see that there was some ecological change, but it wasn't tremendous," Petraglia explains. "There wasn't a complete resculpting of landscapes. It was the kind of landscape that hunter-gatherers could easily readapt to."

Petraglia cautions that a Yellowstone VEI 8 today would initiate a reign of volcanic terror. A big crater in the middle of the United States and a suffocating shower of ash would devastate the nation. The lives of hundreds of millions of people, unable to simply uproot and move elsewhere like their distant forerunners, would be destroyed. Like the Taupo supereruption 26,500 years ago or Yellowstone's first VEI 8, the nation's volcanic Armageddon may come about in multiple phases,[49] with pauses of hours, days, weeks, or years in between, causing a grim period of time in which no one can be certain when it will be safe to clear up the wreckage and start to rebuild.

The specific odds of another Yellowstone supereruption are unknown. It has too few VEI 8 eruptions to plot them on a graph and identify recurring patterns. But VEI 8 eruptions anywhere in the world are incredibly infrequent. The Yellowstone of today is also relatively tranquil, giving no indications of wanting to pay homage to its past magmatic masterpieces.

And as its hot spot moves on, and its chances of erupting gradually fade, I'd argue that Yellowstone is not particularly likely to terrorize you, or your children or grandchildren, on any given day. No matter how you look at it, Yellowstone is not a doomsday device to be feared; it's a natural wonderland to be enjoyed. And while millions of visitors every year revel in its gallery of marvels, researchers will keep listening to the music of the abyss, watch the park metamorphose from above, and peer into its past in order to comprehend its present.

III

THE GREAT INK WELL

Tanzania, 2005. An international team of scientists and a handful of local Maasai guides crammed into two Land Rovers and sped across the ground in pitch-black darkness. The hills and valleys around them had dissolved into shadow, with the headlights of the cars illuminating specks of desert brown and flecks of green. At midnight, they reached the foot of a volcano that rose sharply into the sky. Gearing up and grabbing their gas-trapping equipment, they began to ascend. Huge cracks and sudden drops remained almost imperceptible as they carefully navigated the steep rocky slopes and made their way to the summit, arriving just as the Sun emerged, its light pooling onto the craggy rocks under their dusty shoes.

"I didn't want to look at a lot of pictures beforehand, because I wanted to be surprised," says Tobias Fischer, a volcanologist and geochemist at the University of New Mexico. This was a show of remarkable restraint. The moment you reach 9,718 feet above sea level and crest the volcano's outer rim, you are greeted with a landscape that seems to have been transported from another universe. "When you pop over the edge, it's just like, wow, this is completely different from anything I've ever seen," he says. "It's just so strange."

Fischer and his colleagues were greeted with the site of lava bubbling about in a crater, but it appeared to have been polluted by ink.

Gone were the classic reds, oranges, and yellows of the rest of Earth's lava. Instead, black fluids flew into the air and across the crater floor, quickly cooling, mingling with the atmosphere and turning a silvery white. Hornitos, 50-foot-high gnarly monochrome cones that look like an inebriated demon's attempt at pottery, squirted out black lava in all directions from their multitude of tubes and holes. Classic lava can move at a range of speeds, but on a flat surface, you can outwalk most lava flows. Despite resembling motor oil, this lava was more fluid than water, gushing about as if late for a job interview. And it was cold, too—well, by lava's standards. The basalt shooting out of Kīlauea was cooking at 1,800 degrees Fahrenheit. This lava, fresh out of the pit, was erupting at 900 degrees, a mere five times hotter than the water boiling in your kettle. Sometimes it can flow out of the hornitos and cool so quickly that it freezes in mid-air, creating crepuscular cascades seemingly stuck in time.

As the scientists camped at the summit and held their gas-capturing tubes and bottles over the volcano's vents and cracks, they watched the black soup spurt out of hornitos and zip about just a few tens of feet away. "It's incredible how fast it moved," says Fischer, shaking his head and reliving his disbelief. "It was scary! We didn't know what it was going to do or how close we could get." At one point, "it carved its own channel, it flowed over the edge of the caldera, or crater rim, and then it lit the vegetation on fire." During the night, a lava flow just barely illuminated by a dull red glow suddenly made its way downhill into their camp's kitchen area, forcing everyone to speedily pack up their things and move elsewhere.

Most magma or lava has a fair amount of silica in it, a combination of one silicon atom and two oxygen atoms. Silica likes joining up into long chains that form a sort of skeleton for lava, and the more extensive silica chains you have, the gloopier and more tar-like the lava is. Basalt, like in Yellowstone's large and deep reservoir, has a low amount of silica, so it's fairly fluid. Rhyolite, the stuff in Yellowstone's small and shallow reservoir, is more than two-thirds silica, making it exceedingly gloopy.

This volcano's black lava, though, has so little silica that it runs faster than a hay-fever sufferer's nose. Known as carbonatite, it's packed with (among other things) sodium and calcium carbonates,[1] the latter of which you know as limestone or the chalky stuff that builds up around your water tap. Naturally, the question scientists want to know the answer to is: What the hell? Where did it come from? It's not like we have anything to compare it to, because this mountain is the only volcano on the planet that is erupting carbonatite lava today.

There is no debate. Ol Doinyo Lengai isn't just the world's strangest volcano—it is one of the most peculiar volcanoes in the entire solar system. It is a peculiar expression of the volcano's status: as a cog in a far larger machine, one that is transforming the earth through the planet's most spectacular display of geologic creativity. Something remarkable is stirring beneath East Africa, producing volcanoes and sustaining life that is nothing less than alien.

KNOWN BY THE Maasai people for countless generations, this volcano had long been spoken about by Arab traders too. In an 1870 account, one such trader, by the name of Sadi, was reported[2] to have described the summit in fantastical terms: "one moment it is yellow, like gold; the next white, like silver; and again, black." The first time Ol Doinyo Lengai appeared on a map was in 1855, put together by two missionaries. One of the names they had given it was "Snow Mountain." What they were seeing wasn't snow, though, but the cooled black lava succumbing to the humid air.

"Engai" also appeared on their labels. To the Maasai, that word is synonymous with both "god" and "rain"; Ol Doinyo Lengai, then, means "Mountain of God."

Explorers from the West began summiting the volcano in the early twentieth century, spying rapidly flowing mud that appeared to effloresce into a white salt-like layer. They were describing the volcano's carbonatite lava, a sight so bizarre that they thought it was anything but lava. It wasn't until 1960 that scientists first clambered down into the summit's craters, back when this part of modern-day Tanzania was still named Tanganyika and was still a year away from gaining independence from the United Kingdom. John Barry Dawson, a scientist of the Geological Survey of Tanganyika, scooped up some of the black lava and in a groundbreaking 1962 study[3] revealed that it was a real freak of nature, a concoction of calcium, sodium, potassium, water, and carbon dioxide.

Ol Doinyo Lengai isn't just a chemical oddity. It's also suffering from an identity crisis. This baby volcano, just 370,000 years old, is remarkably steep. You can only build tall, steeply sloped volcanoes if you have a very thick, gloopy magma exploding out of a vent and tumbling down onto the surrounding landscape, piling up to form a steep-sided cone. But this mountain's black lava spreads out too far and too thin to make a conical mountain. Fortunately, much of the volcano's magma isn't carbonatite, but stuff that has a somewhat decent amount of those skeletal silica chains, keeping it gloopy enough to build a steep mountain.

In other words, this volcano is able to erupt two types of strange magma: ultra-runny and pretty gloopy. Gloopier magma is able to trap a lot of gases and stop them calmly bubbling out at the surface. Pop a hole in the cap of a gassy volcano and all those confined vapors violently expand, causing a very angry explosion. That means the gloopier, silica-containing magma creates explosive eruptions, and the black magma expresses itself through lava rivers, ponds, and fountains.

Ol Doinyo Lengai's explosive eruptions, including a recent one in 2007, are quite something: they are accompanied by loud blasts, huge plumes of ash, crackling air, and sometimes startling displays of lightning and thunder. The summit is often partly or completely redecorated each time there is a large blast, with a vast amount of the rock obliterated in place of a new crater. After each major explosive eruption, the crater then fills with black lava. Hornitos, which build themselves over an active pool of lava sitting just below the ground, also begin to grow. The amount of black lava that emerges in one sitting varies, but it's always fairly impressive.[4,5] In 2006, in a mere fortnight, it spewed enough to fill up more than 350 Olympic-size swimming pools.

The volcano's surroundings are also quite odd. A stone's throw from the flanks of the volcano, not far from where wildebeest graze on grassy plains, is Lake Natron. Thanks to the same strange chemicals that Ol Doinyo Lengai regurgitates, this lake is incredibly alkaline—the opposite of acidic—making it a highly caustic environment, enough to eat away at various textiles and plastics. A range of animals aren't bothered by this, mind you: from algae to tilapia fish to flocks of flamingos, the lake is home to many. But when animals perish in the lake, they don't always rot away in the beating Sun. The sodium carbonate in the lake can preserve their bodies by drying them up. Mummification is the appropriate term here: sodium carbonate was used by ancient Egyptians in their mummification processes.[6]

But Ol Doinyo Lengai is unquestionably the region's outlandish zenith. Matt Genge, who studies the shiny bits in volcanic rocks and meteorites at Imperial College London, has visited it three times. "It's a super weird volcano," he says. "When you first hear about carbonatites, you go, so it's kind of like a liquid limestone? And then you find out there's only one erupting." There are a few hundred extinct carbonatite volcanoes around the world, either protruding from the planet's surface or buried under eons' worth of geologic deposits. But this quantity pales in comparison to the hundreds of thousands of "regular" volcanoes that are active, dormant, or extinct on Earth. And there don't appear to be carbonatite volcanoes on other planets. All things considered, it is entirely serendipitous that humanity is around to witness an erupting carbonatite volcano.

The harsh incline makes any ascent rife with risks, but the serious dangers remain at the top. "It is a scary, scary place," Genge stresses. "The crater rim is maybe one to two feet across. And on either side, you've got these steep slopes." Fall down on one side and you'll land in the lava-filled crater. "And on the outside, it just goes all the way down to the bottom of Lengai." A tightrope, of sorts, where you have death by liquid fire to the left or death by blunt force trauma on the right. "I almost went along on my hands and knees," he says.

Gravitationally assisted demises aside, there is a variety of other ways that Ol Doinyo Lengai can kill you.[7] An explosive eruption at the summit would do the trick; avalanches of debris have rocketed downslope for many millennia; minor pyroclastic flows and surges have infrequently bathed the flanks in fiery gas and volcanic matter; the 2007 explosive eruption blanketed the villages of Naiyobi and Kapenjiro (seven and nine miles away from the volcano, respectively) in a couple of inches of ash, the sort capable of burning skin; pools of invisible, odorless carbon dioxide in the summit, sticking around as it is denser than the surrounding atmosphere, can cause asphyxiation if an unlucky creature wanders inside.

The hornitos can ambush you if you aren't careful. "The hornitos start filling up with magma, then the side of the hornito can fail, and all the lava comes rushing out as a sort of tsunami," says Genge. And Angelina Jolie wasn't being careful: at the end of the movie *Tomb Raider: The Cradle of Life*, you can see said raider of tombs climbing up out of a partially collapsed hornito at Ol Doinyo Lengai's summit. "I cannot image a worse place, a more dangerous place to take an A-list star and shoot a movie," he says.

Genge explains that both he and Dawson, on different occasions, have also been caught in the firing line of a hornito sneeze, the ejection of dozens of small droplets of black lava from a hole. It is never pleasant to be hit by flying lava blobs, especially as they scald your skin, but if the notoriously cool black lava hits your clothes, they will likely expunge their heat so rapidly that no bodily harm will result. In fact, despite the risks, those climbing Ol Doinyo Lengai—accompanied by Maasai guides and other Tanzanians with experience scaling various volcanoes, including Kilimanjaro—are able to keep safe. Accidents are few, but remarkable: in 2007, a Maasai porter fell through the crater floor into a scalding pond of black lava. Although badly burned, he ultimately survived his ordeal, perhaps the only time anyone has *fallen into lava* and lived to tell the tale.[8]

Animals aren't as fortunate. After the 2007 explosion, a river of black lava raced down the side of the volcano. Visiting after things had calmed down, Genge found the now-solidified lava flow—and a bone jutting out the top, the remains of a gazelle that was not as agile as it thought.

But few would disagree the danger means the climb isn't worthwhile. Peering into the summit craters, which change from week to week, is a singular experience. Expeditions have seen temporary caves built by black lava. One, found in 1990, was nicknamed Hades. The author of a report[9] on it was wary of getting too close. "The prospect of being gently poached to death should the roof collapse any further did not appeal, and the situation did not encourage prolonged observation," he wrote. Hades's roof was also adorned with eerie, ultra-thin carbonatite stalactites, tinged yellow by sulfur vapors emerging from below. "It is likely that the structures were extremely delicate, although no attempt was made to throw in any missiles to test this theory," the adventurer noted.

Everything up there takes on a strange sort of beauty. On one of Genge's visits, he saw rain percolating down through the crater and reacting with the lava remnants underfoot. "There were these beautiful white flowers, mineral flowers, growing out of cracks on the surface," he says.

Aesthetics aside, scientists wish to venture skyward because they want to work out what is happening at Ol Doinyo Lengai to produce such zany lava. The only way to do that is steal its belches and, more importantly, grab its lava. And getting the latter requires a stunt a little reminiscent of a rather famous one from *Mission: Impossible.*

KATE LAXTON, a PhD student at University College London, originally wanted to do a project on Nyiragongo, a hyperactive lava factory in the Democratic Republic of the Congo. When it seemed unlikely to go ahead, her supervisor instead suggested Ol Doinyo Lengai.

Her background is not in geology, but in environmental conservation. She had long been enthralled by the Serengeti, the savannah plains, and the wildlife roaming about within. It turns out the ash from Ol Doinyo Lengai spreads across the grasslands and fertilizes the soil. "It's just nice to see all the intricate connections rather than just looking at nitty-gritty processes within the volcanic system," she says. "You know, big-picture stuff."

She hadn't given the mechanics of the volcano itself much thought until that point, but as she started to look into it and its weird lava, she thought: What *is* this place? Unlike most volcanoes, this one continually changes its appearance while engaging in a decades-long cycle culminating in a big explosion.

All volcanoes are unique. They are a bit like pets, she says: they all have their individual personalities. But Ol Doinyo Lengai stood out from all the rest. Its black lava made it resemble a volcanologic Rorschach test. So many ideas attempting to explain the plumbing system responsible for this equivocating, eccentric volcano had been put forward, but no one had been able to pin things down. Scientific papers often contain interpretative diagrams designed to illustrate what the authors suspect is happening inside a volcano. But many of the sketches of Ol Doinyo Lengai's magmatic pipes are filled with literal question marks, a reflection of the scientific community's level of befuddlement.

Laxton wasn't the only one to be hypnotized by this Tanzanian troublemaker. Emma Liu, a volcanologist at University College London, has prodded volcanoes all over the world. She has spent half of the past two or three years abroad on fieldwork, from Papua New Guinea to Chile to the South Sandwich Islands just shy of Antarctica. She was at Kīlauea in 2018 with Emily Mason, sampling gases from the hyperactive fissure 8. The moment I told her about my conversations with Mason, she immediately says: "Have you heard about the lava fishing?"

I had not. It turns out that Frank Trusdell, a research geologist at the US Geological Survey, used to grab samples of lava from the channel coming off fissure 8 with, well, a standard-issue fishing rod. "He walked right up to the edge, fished, reeled it in, wiggled it around for a bit, pulled it back, and then somebody poured a bottle of water over it to quench it," she recalls.

When it comes to Ol Doinyo Lengai, chuck almost everything you know about volcanoes out of the window, she says. "Everything we know to look for in terms of patterns and sequences don't apply to Lengai."

Wanting to unravel a handful of the volcano's many perplexing idiosyncrasies, Laxton figured that one of the best ways to do so would be to grab samples of gases flying out of Ol Doinyo Lengai and scoop up some of its carbonatite lava. In 2018, she headed to Tanzania for a bit of reconnaissance. "I had no idea what I was getting into," she says. She had never been to Africa before, nor had she ever conducted a remote expedition. "It was a good idea to go and see what I was up against." It turned out that one huge obstacle was, as ever, red tape. You need a lot of contacts and permissions from all kinds of local and regional authorities to do work like this in Tanzania. You also need to know people on the ground beforehand in order to travel to the site to get to know people and request permissions in the first place. Tackling that bureaucratic *ouroboros* was decidedly not fun.

More memorable, and more physically exhausting, was the hike to the summit. "The worst of it is that you start at midnight, so you're already tired," she says. "You climb up at night so you can't see anything but this patch of light. For most people that's helpful because if you're scared of heights you can't see the big ravines either side of the path. A lot of people find the descent a lot worse because you're actually facing downslope and you can see how steep it is, you can see all the ways in which you could have tripped on the way up. There's nothing to stop you rolling all the way down." Why do the guides start everyone climbing to the summit at midnight? It's for the tourists, she says: seeing the sunrise as you crest the peak is pretty special, so long as you don't suffer from acrophobia.

At the top, she found that you are completely exposed to the elements. She stayed overnight at the summit in a tent. Sloshing lava was audible, but the crater wasn't exactly putting on a show. Hopes were nevertheless high that, on her return, there would be plenty of lava to collect.

Armed with local connections, gaining the support of the Geological Survey of Tanzania and the country's University of Dar Es Salaam, and understanding the conditions at the top, Laxton, Liu, and other colleagues returned in 2019. "It became apparent as soon as we arrived in the region that this volcano is respected," says Liu. They had to be really sensitive to how the Maasai felt about it all. "Just camping on the top for example could have caused tension, so we had to go about everything really carefully."

The team saw Maasai praying at the south crater at one point. "We were there during a period of drought, so they were up there to pray for rain. And it did," says Laxton. The entire Maasai culture revolves around the volcano. Pilgrimages from all over East Africa are made to the volcano, to pay respects. "Depending on what they do, it either angers Engai or pleases Engai," she says. Explosions are generally seen as bad signs.

Naibala, a local and, since 2009, a guide to the volcano based in the nearby village of Engare Sero, says "the mountain Ol Doinyo Lengai is very important for us as a Maasai community, since we believe in nature." Rain is often prayed for, but Engai is a fairly multipurpose deity. Women who are having difficulty conceiving "can go there to pray in order to have kids," he explains, adding that the Maasai community sometimes performs animal sacrifices to Engai, both at the base of the volcano and at the summit. When an important prayer is needed, village elders appoint four men and four women to act as the community's envoys, to head to the volcano and ask the god for a helping hand.

At the same time, Naibala says, the community is aware of the scientific work being conducted on the volcano, with many cognizant of the fact that the strange lava it produces means something unusual is happening here. As long as the visiting scientists make sure to be respectful, the Maasai are happy to gain a better understanding of the volcano they hold in such high regard.

"There was an element of just acknowledging the hierarchy there," says Liu. "And then there were the obvious financial transactions that show that you respect that it's their land." Government permits are not recognized there, so everyone needs to pay for access to the region just like any tourist would—although Laxton, as the group's leader and a student, had her fee waived. When it came to conveying respect, it wasn't so much what you said that mattered, Liu explains; it was your actions, which needed to clearly demonstrate that you were visitors on their territory.

With everything green-lit, Laxton and her team clambered to the summit of the volcano and were thrilled to discover that lava was dancing about with vim and vigor in the summit crater. Camping in an elevated, plant-adorned, saddle-shaped section, they could easily see the often-hourly changes in the crater down below. Every day, at first light, they photographed the morphing crater floor from a safe distance. The lava wasn't always black. When the volcano oozed gassier lava, it took on a shiny brown tincture. On the last day, they saw bright orange lava flows. "After the rain," says Laxton, "it turns blue and green."

But the carbonatite lava wasn't the strangest aspect of their nine-day sojourn at the summit. It was the diversity of wildlife that took them by surprise. You wouldn't expect such a mercurial, death-enabling locale to feature a cornucopia of critters, but up there they saw all sorts, including baboons, cheetahs, and antelope-esque elands. It seems that, despite the volcanism, plentiful vegetation, cooler air temperatures, and salt deposits are pretty amenable to life. "I think the eland come up to the crater to lick the rocks," she says. "It's kind of cute." Back during the 2018 reconnaissance trip, "the crater was completely full of butterflies, hundreds and hundreds of these white and yellow butterflies, which was amazing to see," says Laxton.

Liu remembers feeling somewhat engulfed by the wildlife. "I was convinced that we had leopards over there," she says. "And it turned out they were small dassies, these little rodents. But in my sleep-deprived state I'd made them into these giant leopards." They did eventually see some leopards, she hastens to add, so she wasn't going completely mad.

You might think that the volcanic murmuring in the middle of the night would be at least marginally unnerving too, but you would be wrong. "The sounds from the volcano? They were really comforting," says Liu. "If I could hear it gurgling, I was like, ahh, excellent, lovely, that makes me comfortable." The disembodied screeching of the baboons, though, was anything but. These marauding primates also meant that answering nature's call was sometimes a bit of a gamble. Toilets do not naturally occur atop Ol Doinyo Lengai, so your only option is to drop your pants and relieve yourself over the edge of the volcano. "But if you go over the edge and you're in a vulnerable position, and you see a whole troop of baboons coming up toward you, it's . . . not something you can prepare for," Laxton says. What else is there to do in such a situation? Abort, "pull up, and run."

Despite the threat of those errant baboons, the team still had to study the volcano. They had a range of objectives, including measuring volcanic gases on the flanks of the volcano and in the crater, mapping the crater using a drone, charting the heat output of the volcano's upper segment, and so on, sometimes with the technological assistance of other universities or institutions. But obtaining samples of Ol Doinyo Lengai's lava got top billing. Very few fresh samples of the black lava had been obtained in the past, because one does not simply walk into Ol Doinyo Lengai's crater with a shovel or, if you fancy, a fishing rod. It's too dangerous and unpredictable an environment. The crater also happened to be sitting at the bottom of an incredibly steep crater wall—a drop of nearly 330 feet, enough to fit the Statue of Liberty inside with room to spare.

Laxton happens to be an experienced climber. She met her partner while climbing, and she is a huge fan of practical, mechanical thinking. "Technology always seems to crap out in front of me," she says. A drone may work, but the wind at the summit was so strong that there was a risk it would be blown into the crater and lost forever. So, she thought, why not use her climbing experience here? Was there a way to get someone or something to rappel into the crater and close enough to the fresh lava to scoop up some samples?

Lava is certainly rather hot, but we have invented plenty of things that lava cannot melt. Laxton had used nothing less than a hammer to scoop up lava into a bucket in Hawai'i back in 2013. She has seen people on YouTube playing with lava with a spatula without wearing any protective gear. "Scooping it up is the easy bit," she says. Getting the scoop to the lava? That's the problem when the lava is 23 stories below you.

Laxton had a plan. But like the best heists, she required a crew to pull it off.

She'd need another damn good climber to help out. That came in the form of Arno, who was spotted dangling from a large building near her university's campus. They had coffee, she explained her mission, and by the end of it, this professionally certified master of ropes and pulleys was heading to Tanzania. She would also need some customized gear to enact her plan. Fortunately, a climbing equipment company in Wales produced a bespoke, lightweight, highly adaptable pulley that was perfect. After testing it out at a Welsh quarry, she knew she had the tools required. Laxton also enlisted the help of another PhD student in France. They had previously used plastic clothes-hanging lines to lower a GoPro into the mouth of a volcano; a few tweaks, Laxton realized, and this system could be adapted for some lava thievery. And, most important of all, Tanzanian porters would make sure they kept alive and well at the summit by relaying food and water to the scientists every two days.

Her crew was assembled. The heist was on.

In 2019, at the volcano's summit, they set up anchor points around the circumference of the crater rim and joined up two of them, effectively creating a tightrope-like pulley system spanning the crater from one side to the other. Too hazardous to send a human down on a rope, Laxton instead opted for scooping devices attached to a pulley. These scoops would be winched out onto the tightrope before being carefully lowered down into a pond of black lava as binocular-wielding helpers looked on and guided the descent. Ideally, while this was taking place, no one would fall into the volcano's maw.

Those scooping devices? It turns out that there is nothing better to hold samples of black lava than stainless-steel cocktail shakers and wine measures. Laxton had previously exposed this game-changing kit to a blowtorch to check whether it could withstand carbonatitic lava temperatures. The containers didn't even flinch: stainless-steel has a melting point three times that of black lava.

The stakes were high; much of Laxton's PhD thesis was riding on the success of this work. Two containers were chosen to be the inaugural lava-dipping pioneers. Carefully winding them out over the crater, making sure their alignment was perfect, and winching them down, they slowly but surely fell under the benighted pool of lava. Arno shouted "Pull!" and Laxton ran downslope, using gravity-assisted momentum to winch the cups back up. But she yanked on the pulley too vigorously. The friction caused the rope to splinter, then break; the cups fell into the lava, their home forevermore.

This was thankfully their one and only scoop failure. The rest of the cocktail shakers and wine measures refused to join their brethren in the flames, and the team watched as six fresh samples of black lava journeyed back to them, whereupon they were air-dried like slices of geologic jerky to preserve their minerals and pristine textures. Although other research aspects of the mission were a bit more hit-and-miss, the lava heist was an indisputable success. By this mission's end, everyone's personal hygiene went south. But their triumph, one achieved thanks to Laxton's preponderance of planning, community engagement, and ingenious low-tech rig, made it all worthwhile.

Now, says Laxton, comes the hardest part: decoding the lava. Previous samples had been taken from much later in the volcano's effusive-to-explosive eruptive cycle, says Liu. Theirs had been obtained nearer to the start of the cycle, relatively soon after a big explosion. Seeing how the chemistry changes from one to the other may tell scientists why this cycle exists at all.

Not all cycles are the same, however. No one expects any volcano to erupt according to a rigid schedule, and Ol Doinyo Lengai is no exception.

Back in 2007, an explosive eruption rocked the summit, and romance was canceled. "It happened to be that my former student was on a honeymoon traveling through Africa and he happened to be at Ol Doinyo Lengai, driving by," says Fischer. The vagabond pupil telephoned Fischer, telling him of the eruption and asking for some instruments to be shipped out there so he could grab some data from it. It isn't clear how happy his partner was with this arrangement, but well, that's science for you.

In 2009, Fischer and his student went back to the volcano to grab some more samples. The chemistry of the lava suggests that a cache of gloopy magma was injected into the volcano, topping up its supply. Thanks to that spicy, gassy kick, the pressure of the pile of gloopy magma rose, and the rock keeping it confined belowground could no longer contain it. Boom: a hole was carved out at the summit, and the cycle began anew, with black lava pouring into the new crater ever since.

There is still not much data to go on, and only a handful of eruptive cycles have been documented at Ol Doinyo Lengai. But based on the previous few, the explosive climax in 2007 was a little earlier than expected. Perhaps because of that injection of magma, says Fischer, "it got a little kick in the ass."

LAXTON AND HER TEAM are just some of the researchers actively studying Ol Doinyo Lengai. It is so difficult to conduct successful fieldwork at the volcano that any and all data are vital. Get enough of it, and you can solve a tiny piece of the puzzle. And over the past few decades, enough pieces have been collected that one of the overarching questions about the volcano—namely, why the heck does it make lava that is the molten equivalent of limescale—has something of an answer.

Lava is just magma aboveground, and the ingredients for that magma and the gases trapped inside it come from a source. If that source is identified, then you know what made the volcano. So grab some lava and its gases, and you can solve the puzzle.

Karim Mtili is working on his master's degree at the University of Dar Es Salaam. To say he knows his stuff is an understatement: at the time of writing he is yet to finish his degree, and the university has already offered him a permanent lectureship. He is keen to hunt down helium in the region. Helium belongs to a category known as noble gases, a reclusive bunch that are so chemically languid that they rarely react with other substances. That, Mtili tells me, means that helium can emerge from hundreds of miles belowground and reach the surface in the same form as it was when it began its journey.

There are two atomically stable flavors of helium: helium-3 and helium-4, with the latter being a smidge heavier thanks to the extra neutron sitting in its nucleus. Helium-3 is what is known as primordial. It was trapped inside Earth during the planet's formation billions of years ago, where it is now stored below the crust in the mantle. Helium-4 is the new kid on the block, a by-product of the radioactive decay of uranium and thorium in the crust. Volcanism lets off helium gas from time to time as its magmatic broth rises. If you have a much higher proportion of helium-3 coming out of the broth, you know that the magma was cooked up in the mantle. More helium-4 suggests that a significant part of the cooking process happened in the crust.

Helium is just one gas that can be used to trace the source of a volcano's fire. Fischer managed to bag a lot of excellent samples of various volcanic gases on his 2005 trip. He helped show that the boatload of carbon composing Ol Doinyo Lengai's magma was derived from the mantle. In other words, the black lava has its origins not close to the surface, but far belowground.

Another decade of work in the region, by Fischer and others in the community, built off initial work like this, culminated in a 2020 model[10] attempting to explain what makes this volcano tick.

The lithosphere—the crust and upper mantle sandwich that makes up the skin of the world—isn't the same everywhere. Ol Doinyo Lengai, and a trail of dead carbonatite volcanoes, sit on a boundary. To the east, you have a younger, thinner lithosphere. To the west, there is a big chunk of mangled up continental rocks, a 3-billion-year-old or older lump named the Tanzanian craton. Over its lengthy history, mantle plumes have risen from Tartarus, tickling the underbelly of the craton and supplying it with plenty of solid carbon. (This carbon, by the way, is likely where diamonds come from. They aren't made from compressed coal up in the crust—that's a myth. Diamonds were made in the mantle and were jettisoned up to the surface long ago by volcanic pipes at speeds ranging from 100 miles per hour to 1,300 miles per hour.)

The Tanzanian craton is so thick that its carbon-rich base sits really deep down within the mantle. This carbon is under so much pressure that, try as it might, it cannot melt. But this deep carbon is perfectly happy to slide up along the boundary between the craton and that thinner, shallower lithosphere to the east. As this carbon concoction rises toward the upper mantle, it decompresses, allowing it to melt. These melts have no trouble punching through the thin lithosphere, so they shoot up toward the crust, where they pool and brew for a little while.

Now you have carbon-rich melts and silica-rich melts. They do not get along. "It's like oil and water," says Liu. "They all keep separate. And it's kind of the same for magmas. If they're oversaturated in a particular component, like carbon, carbonates, it will be thermodynamically favorable to separate them into two fluids." At a critical point, the silica-rich melt tells the carbon-rich melt to get stuffed, and the carbon-rich melt leaves, becoming carbonatite magma. Being less dense than the silica-rich melt, it sits on top of it like a cap.

Many volcanoes around the world emit carbon dioxide. Ol Doinyo Lengai erupts *a lot* of carbon dioxide. This is how volcanoes normally get rid of their carbon. Under low pressures, carbonatite breaks down. If you take a limestone and try to melt it just below the ground, it falls apart and you get carbon dioxide and calcium oxide. So how does Ol Doinyo Lengai manage to erupt black carbonatite lava? Like many recipes needing a little something extra to make it stand out, all you need is a little salt—well, the element sodium, to be more accurate.

Coming along for the ride, this sodium bonds with our molten limescale magma. Sodium turns the carbonatite into a *natrocarbonatite*, which is entirely capable of remaining stable as a liquid at the surface. And that is how some scientists suspect Ol Doinyo Lengai makes its ludicrous lava.

Much progress has been made on the question of Ol Doinyo Lengai's strange chemistry. But a second overarching question remains wide open: Why is this the *only* volcano in the world that erupts this aberrant lava? What makes Ol Doinyo Lengai so special?

The boundary between the Tanzanian craton and the thin lithosphere to the east is huge, so why is Ol Doinyo Lengai the only one taking advantage of its environment? Fischer's answer reflects everyone else's that I spoke to. "I don't really have a good answer for that," he says, shrugging. It could be that some volcanoes nearby have erupted carbonatites in recent geologic history, but the evidence has been buried by more normal lava flows. No one knows for sure.

Answering these questions won't just reveal more of Ol Doinyo Lengai's secrets. The volcano is, as Fischer puts it, a gateway to the mantle. The mantle makes up 84 percent of the volume of the entire planet, so understanding how it works, and how it affects what happens on the planet's surface, is of paramount scientific importance. Rather inconveniently, the mantle is so deep below us that it remains inaccessible to most of our technological tools. But here, on either side of the thick Tanzanian craton, something odd is transpiring. The mantle is remarkably close to the surface. Far too close, in fact.

Remember those mantle plumes that gave Ol Doinyo Lengai its solid carbon fuel over billions of years? It seems a far more colossal sibling has been at work beneath the entire region over the past few tens of millions of years, causing this corner of Africa to bulge up, crack, and rip apart, giving rise to an entire family of strange volcanoes stretching for thousands of miles. Ol Doinyo Lengai may be a charmingly rogue mountain, but in the grand scheme of things it is a single thread in the planet's most baroque volcanic tapestry.

EAST AFRICA IS disintegrating.

The Arabian Peninsula—the big block of land containing Saudi Arabia, Yemen, Oman, the United Arab Emirates, Qatar, Bahrain, and Kuwait—is part of the Arabian tectonic plate. Many millions of years ago, it began drifting away from the African plate, which includes most of the African continent and a decent portion of the Atlantic Ocean.

The African plate itself is now breaking apart. Along a broad swath of land thousands of miles long, from the Red Sea sandwiched between the Arabian Peninsula and East Africa right down to Mozambique, the African plate is roughly split into two plates: the Nubian plate sits to the northwest and the Somalian plate to the southeast. Both plates are moving in opposite directions. This tectonic divorce is known as the East African Rift. East Africa is, by just a few inches every decade, breaking apart.

Christopher Jackson, an expert in geologic basins at Imperial College London, and Lucía Pérez-Díaz, a breaker and shuffler of tectonic plates at the University of Oxford, walk me through the rift's evolution and intricacies. This great tear in East Africa began somewhere around the African–Arabian plate boundary. Up at the top, you can find a triple junction, where you have the Arabian, Nubian, and Somalian plates all moving in different directions. "And what's in the middle of this triple junction is not, like, the Eye of Sauron or something, or where the kraken is going to come up or anything like that," Jackson chuckles. Phew.

Instead, we have the Afar Depression, a huge topographic bowl in Ethiopia. The lithosphere here has been stretched thin by the outward movement of those three tectonic plates, like hands pulling apart fresh pizza dough. Volcanoes aplenty have erupted here, both onto and into the crust, adding weight to it and causing it to sag downward into the depression we see today.

From here, the rift then gradually unzippered toward the southwest. Around 25 million years ago, it encountered the chunky, unassailable Tanzanian craton. Deciding it would take too much effort to shatter, the rift split in two, going around it to the west, from Uganda to Malawi, and to the east, through Ethiopia, Kenya, and into Tanzania. Both branches have volcanoes, but the west is mostly known for its pronounced earthquakes while the east is renowned for its volcanism. Ol Doinyo Lengai sits close to the southern extremity of the eastern rift section.

The culprit for this act of continental sundering isn't trying to hide. The chemistry of the lavas erupted by the dozens of active volcanoes all along the East African Rift clearly shows that many get their ingredients from the mantle. Two huge domes, one rising above Ethiopia and another atop Kenya, are thought to be manifestations of mantle plumes effectively pushing up the crust. Scientists more or less agree that a series of huge mantle plumes or one large supercharged mantle plume—the so-called African Superswell—sits under East Africa. This huge mass of material is so buoyant that it elevates the crust above it; so hot that it erodes and thins the crust as it does so; so mobile that it helps pull the African plate apart in two different directions. In doing so, it has authored a volcanic tour de force.

"If you want to find volcanoes, go to the East African Rift because it's kind of beautifully set up," says Jackson. The crust is stretching, the lower mantle moves up, it decompresses and melts, producing lots of weird magmas, and the faults and cracks in the crust allow volcanoes to pop up all over the place.

You can find volcanic environments elsewhere on the planet that have some similarities to the East African Rift. "But it's more extreme here, and it's really long, right? It's huge," says Fischer, gesticulating excitedly. "It's a very majestic place, it makes you feel like, oh man this has been around for a long time, but it's also quite dynamic—you have volcanoes that erupt there, you have crazy volcanoes."

Ol Doinyo Lengai is easily the oddest of the bunch. But Ethiopia's Erta Ale shield volcano is extraordinary too, featuring one and sometimes two long-lived, scorching lava lakes at its summit. Nyiragongo, over in the Democratic Republic of the Congo, is a vertiginous volcano full of the potassium, sodium, and calcium compounds that make up Ol Doinyo Lengai. It doesn't erupt black lava, but its red-hot lava is so fluid that it can move at more than 60 miles per hour, enough to catch up to a speeding car. Although eruptions at these fiery mountains can be and have proven to be deadly, this volcanism is also extremely valuable: it digs up treasure.

Helium is one of the best coolants money can buy. In liquid form, it is able to keep expensive medical and scientific equipment from overheating and blowing up. Mtili has explored the chasms around the Tanzanian craton, where radioactive decay produces a decent amount of it. Volcanoes cough it up to the surface, where it sometimes gets trapped in rocky seals from which it can be mined.

Carbon dioxide (CO_2), the stuff you exhale, may not sound particularly useful. But soft drink manufacturers disagree. Tanzania's Rungwe Volcanic Province, a collection of relatively tranquil volcanoes, is still sneezing a huge volume of CO_2 into the sky. "The Coca-Cola Company use it to source their carbon dioxide," says Mtili, the very bubbles that fizz in your soda.

Mtili and his colleagues recently went to the province in order to find where this carbon dioxide was coming from. "The gas seepage was occurring along the riverbed, so there were a lot of bubbles coming out. And when we passed through it, when we found the coordinates, we were so excited, all of us, we rushed in to set up our sampling gear. And suddenly, we felt dizzy," he says, laughing. They continued trying to set up the equipment as their heads continued to lighten, when they had an epiphany. "Oh, this is carbon dioxide guys!" he recalls shouting. They suddenly noticed all the dead birds and snakes around the seepage site that had suffocated under the veil of transparent volcanic gas. Mtili and company awkwardly escaped death's gassy grip and caught their breath away from the riverbed's wretched reach. "We forgot about the basic biology of carbon dioxide," he says, chuckling away. "We almost fainted, all of us."

Despite the occasional dance with death, Mtili says he never gets tired of studying the East African Rift, and not just because the science is so rewarding. The cultures surrounding its volcanoes are just as intriguing. In one spot, the rift's volcanism causes a salty brine to sneak up to the surface, where it is at the very least hot and sometimes scalding. "The people believe that it has some healing kind of thing, they take a bath, some of them use it to make salt for home use and the business," he explains. "And their belief is that there is some kind of supernatural deity that brings them that water." When the deity is angry, the temperature increases.

There is a clear generational gap when it comes to the scientific understanding of these volcanoes. "The older generations—the seventies, eighties, old guys—they still hold on, they're really conservative with their beliefs," says Mtili. The younger generations are given a scientific education on what their volcanoes are doing, and why. But although the science is accepted, many also maintain the belief there is also something supernatural going on. *Trust me, I know*, they often say to Mtili. "It's a really interesting coexistence of knowledge. The younger generation hold on to their ancestral knowledge, but at the same time, they understand the science of what's happening."

In any event, local villagers are almost always interested in understanding more about the volcanoes they revere. They get involved, and when scientists come to do fieldwork, they have an encyclopedic knowledge of the volcano and where best to find what. "They know everything that's going on," says Mtili, including the sort of research that's being conducted. They help scientists find the best spots to conduct their work.

For a long time, foreign scientists—from both the West and elsewhere in Africa—have "used the locals just as carriers, carrying the equipment and so on, which is really not good," he tells me. But things are improving. I ask him about the way Laxton's research was conducted in coordination with local researchers and those living near Ol Doinyo Lengai. This "was miles better, miles better, yeah," he says.

"The more we get the word out there about how much locals want to get involved in this research . . ." he says, before pausing for thought. "They're not stupid people, you know? They have common sense too. When you treat them with respect and try to explain to them what you're actually doing, try to show some appreciation of their cultural knowledge, they are really happy to help."

The work of D. Sarah Stamps, a geophysicist at Virginia Tech, Elifuraha Saria, a geoscientist at Tanzania's Ardhi University, and Kang-Hyeun Ji of South Korea's Korea Institute of Geoscience and Mineral Resources, is another sign that times are changing. They have set up a network of GPS sensors on and around Ol Doinyo Lengai that relay information on the changing shape of the volcano—inflating or deflating, depending on what the magma and its various unguents are doing belowground—in real time back to the researchers. If the volcano is transforming in a way that suggests something wicked is afoot, they can alert local officials.

All the sensors are owned by Ardhi University, and the long-term plan is that they will take control of the entire network. "We're training Tanzanian students, so it's not just going to be foreigners coming in and doing the science and leaving," says Stamps. "We are investing in the education of the country." Their ultimate goal is to keep adding scientists and equipment so that, eventually, they can set up a full-fledged volcano observatory in the region. "We can't wait for that day," she says. "It's going to be great."

THE SCIENCE IS IN: lava, while molten, is bad for life. It flambés it to death. But microbes can thrive in conditions most animals would find abhorrent. The same is true for much of the East African Rift, with watery pools home to all kinds of life sitting between fresh, jagged lava flows still baking underneath a thin crust.

And then you've got Ethiopia's Danakil Depression.

Sitting within the larger Afar Depression near the top of the rift, Danakil is more than 330 feet below sea level. Only a volcanic mound along the coast stops it from being flooded. "But when the sea level was higher" in the distant past, says Jackson, "there used to be marine waters in the Danakil Depression." Ancient coral reefs, marine terraces, and salty minerals stand as memories to the watery land this once was.

To say it is a hostile place today is a gross understatement.[11] This 160-mile-long bowl has daily average temperatures of 94 degrees Fahrenheit, making it one of the most Sun-scorched places on Earth. Rain is a rarity, with only four inches of the good stuff falling on Danakil each year. And within Danakil you can find a volcano named Dallol, flanked by countless hydrothermal ponds and pools. It is a polychromatic hellscape: volcanic chemicals brought up by the East African mantle monster decorate the waters in vibrant greens, luminous oranges, milky whites, and sickly yellows. The air bakes in the stench of rust, eggy sulfur, and acrid chlorine. These viscous vapors bite the inside of your nasal cavities and fire tiny splinters into your lungs if you dare to approach the small mineralized mounds looming over the wheezing pools.

Purificación López-García, a microbial diversity expert at the French National Center for Scientific Research, has worked around Dallol in temperatures exceeding 122 degrees Fahrenheit. You have to drink water constantly and move very slowly so you don't overheat. Working here and in the surrounding region was, she says, "the most extreme experience I had in my life." Felipe Gómez, a microbiologist with the Astrobiology Center in Madrid, Spain, says that Dallol looks very beautiful, but it is incredibly hot and full of death. "Some birds see water and dive down [to drink] and die over there," he says. "It's possible to see some of the small pools surrounded by birds that have died. Yeah, it's horrible."

But, as the axiom goes: where there is water, there is life. As long as there is water and some chemical compounds that can be used to make energy, certain species of microbes—extremophiles—can make Dallol their domicile. Remarkably, some microbes can exist in a combination of two of these environmental extremes, making them polyextremophiles. And showoffs.

The same applies to Dallol, at least up to a point. Hardy microbes have evolved to withstand very high temperatures. Some can live in extremely salty environments too, while others thrive in remarkably acidic environments. "We normally say that pH is zero, because we have no probes adapted to measure a negative pH. In theory, a negative pH doesn't exist," says Gómez. But here, some pools almost certainly have a negative pH. And yet, life finds a way.

Dallol has hyperthermal pools, hypersaline ponds, and hyperacidic puddles. But what makes it such a draw for microbiologists is that, thanks to its anomalous and enormous volcanic fuel source, it has bodies of water that check all three boxes, making it the most extreme environment for life on Earth. Life flits about in some of these watery pools, but others, says Gómez, are completely sterilized. Sometimes the extremes are, well, a little too extreme for life. Sometimes the water in the pools is so bound up with other chemical compounds that not enough is available for microorganisms to use themselves.

"In every extreme environment where I have been, if I found water, I found life," he says, speaking of his adventures to both the North and South Poles, the Atacama Desert, and so on. "The only exception is here at Dallol."

Barbara Cavalazzi, a geobiologist and astrobiologist at the University of Bologna, says that this is the perfect place to study astrobiology. If life is found in these pools, then it is living right at the limits of biology on Earth. If life is absent, then perhaps we have found a set of conditions that prevents any life from existing, a concept that may apply to environments on other worlds. That makes finding where the limit lies in Dallol—arguably the most otherworldly consequence of the East African Rift—a scientific endeavor with profound consequences.

In 2019, Gómez and his colleagues published a paper[12] that looks to have pushed the limits to life even further into the unfathomable beyond. A few years back, while using the Danakil Depression to calibrate a scientific instrument that would ultimately end up on NASA's *Curiosity* rover on Mars, they poked about in the pools around Dallol and found fatty acids, the sort that can be found in cells. While ruling out possible contamination from scientists, tourists, and animal life, they identified minuscule round structures entombed within mineral deposits. They concluded that these were the cells of bacteria.

This was a huge surprise. Life on Earth had never been found in a simultaneously hyperacidic, hyperthermal, and hypersaline environment, the combination seemingly proving too much to handle. And yet, in this pool of death, it appears that something could survive. Stranger still, these sorts of microbes had been found before in other salty environments, but not in this most extreme of settings. Gómez and his team are currently trying to unravel the biophysical mechanisms that permit the microbes' survival, work that is truly at the leading edge of microbiology.

Later that year, López-García and her colleagues published a paper that indirectly ran contrary[13] to this discovery. Looking in similar pools with all three environmental extremes, they came away empty handed. They found that the preponderance of salt and acidity proved to be too disruptive to life. High concentrations of magnesium salts present were capable of shattering chains of molecules and the protective membranes wrapped around cells. "They suck up the water" that cells need, she tells me—a fatal blow for microbes. The team also suggested that tiny grains within the pools certainly resemble cells, but are nothing more than mineralogic mimics that can be misinterpreted as evidence for life.[14]

Gómez remains upbeat and confident. A few years ago, he says, he would not have believed the results of his paper either. But the limits for life seem to only extend further away as each year passes.

What everyone does agree on, though, is that the only reason life can function at all at Dallol is because of the volcanism that permeates

East Africa. Volcanoes provide crucial heat and the flow of chemicals that microbes can use as energy sources. Sprinkle on some water, and you have a home for adventurous little critters.

Volcanoes can be dangerous. "But from a scientific perspective, it's quite the opposite," says Gómez. "The origin of planet Earth is volcanoes." And no one knows for certain, but these sorts of habitats are probably like those that first gave rise to life, from which every living thing today has descended.

IT ISN'T JUST microbial life that the East African Rift's volcanism supports. Humans, millions of which live along it today, have been wandering the region long before recorded history began.

Along the southern shore of Lake Natron, in Ol Doinyo Lengai's shadow, a set of curious footprints in the ashy mud were found by Kongo Sakkae, a Maasai living in Engare Sero, at some undetermined point in the recent past. But shortly after the turn of the millennium, he happened to mention it to a few staff at a nearby ecotourism camp. An American conservationist happened to be staying at the camp at the time. Returning home, he told his friends about those footprints, including Cynthia Liutkus-Pierce, a sedimentologist and paleoenvironmental scientist at Appalachian State University.

"I got the first pictures of the site on April Fool's Day 2008," she tells me. She thought it was a practical joke being played on her, the brand-new faculty member. But her colleague on the ground quickly proved they were real. In short order, she put together a team of experts; they flew out to the site in 2009 to conduct a forensic examination.

The footprints had previously been hidden beneath the soil. The nearby seasonal Engare Sero river ran through the area and managed to chip away just enough of the muddy cap to reveal a handful of the footprints to passersby. When researchers painstakingly dug down into the dirt, they found 400 individual footprints, all made by humans walking through the region anywhere between 5,000 and 19,000 years ago.[15] The footprints were impressed into an ashy, muddy debris flow that had fallen from Ol Doinyo Lengai sometime before these humans had arrived. After they had departed, another avalanche of mud and ash covered up the footprints, protecting them from erosion for thousands of years.

Thanks to the world's strangest volcano, scientists had a window into our species' past, an ancient record of human activity in the region. "I'm not going to lie. The first time I got out of the car and saw it I cried," says Liutkus-Pierce.

Fossils, like preserved bones, tell you about form. Trace fossils, like these footprints, tell you a little about form and a lot about behavior. It was clear that this was a big group of people. For some reason, a very tall man was running on the peripheries. One person had a big deformed toe, while another tiny-footed person walked a bit like a duck, with feet splayed outward. The mix of individuals in the group—mostly women and adolescents, with just a few men—suggests this may have been a party of foragers, looking for food and resources along the lake's shore. This group has echoes in the behavior of the region's still-extant Hadza people. "It's really hard to speculate though," says Liutkus-Pierce.

Such a harsh environment may seem like an odd place for people to wish to visit. But, just like at the volcano's summit, the area's resilient plant life attracts animals, while volcanism powers watery springs. This wasn't just a transient site, says Liutkus-Pierce, but a place of occupation. And older fossil evidence throughout the East African Rift tells us that "it's had plenty of draw for a very long time."

At the time of writing, the oldest known fossil of *Homo sapiens*—that's us—comes from Morocco, a 300,000-year-old skull.[16] But the remains of all kinds of archaic humans, from *Homo sapiens* to our evolutionary cousins and ancestors, have been found all over Africa,[17] from Ethiopia to South Africa. Tanzania has a fair share too: Olduvai Gorge contains nearly 2 million years' worth of human fossils and artifacts, while the Laetoli site, just 60 miles southwest of Engare Sero, has 3.6-million-year-old footprints made by a human ancestor.

Many places on the continent can make a claim to be the Cradle of Humankind, but it's starting to look like much of Africa was one vast nursery for our species. The East African Rift was certainly part of that nursery—a volcanic marvel, inhospitable and habitable all at once. "As Tanzanians, we're always really proud of knowing that the science supports that part of the East African Rift was where the first humans lived," says Mtili.

A waterlogged volcanic environment, not unlike those scattered along this titanic geologic tear, may have overseen the origins of all life on Earth. Recognizing that the same sort of extraordinary empire of volcanoes could have looked on as our own species took its very first steps into the light does nothing less than stir my soul.

ONE DAY, Ol Doinyo Lengai will enter an eternal sleep. The god within will vanish. The black lava will flow no more. Perhaps another volcano able to perform the same magic trick will appear somewhere to the south. Perhaps not. But, for millions of years more, the mantle below will continue to rise and churn. East Africa will keep unzipping.

The East African Rift could fail. The eastern branch has lost some of its oomph over geologic time; today, it's tearing, but slowly. The plentiful earthquakes on the western branch, though, suggest that section has a little more vigor. Eventually, the eastern branch could stop rifting altogether while the western branch continues apace. There are so many variables at play that the fate of the rift remains fuzzy. No amount of hypothesizing can produce a convincing answer. If the tectonic plates lose their momentum, part or all of the East African Rift could seize up. The scar tissue of failed rifts can be found all over the planet. Earth, easily distracted, doesn't always finish what it starts.

Jackson, peering into the hazy future, remains optimistic about the East African Rift's chances. "It looks like it's still got enough push to kind of get it to tear apart," he says. And what happens if it succeeds?

What is the endgame of eons of piecemeal continental destruction, signposted by undeniably epic, otherworldly volcanism?

Have you ever noticed that the eastern shoreline of South America looks a lot like the mirror image of the western coast of Africa? Look at a world map. They fit so perfectly, right? They resemble matching pieces of a colossal jigsaw puzzle.

This isn't a coincidence. Roughly 140 million years ago,[18] both continents were joined, existing as one single landmass. But powerful forces cast them both apart. The mantle rose. A rift appeared between them. Seawater flooded into the abyss. An ocean was born.

In the southern end of the Red Sea and in the Gulf of Aden, you can find crust—not the sort you find on continents, but the sort you find in oceans. This crust appears when tectonic plates move apart from each other. The mantle rises, melts, and produces somewhat pure strains of magma. It cools into a dense rock that forms the base of oceans. And if you are seeing oceanic crust at the top of the East African Rift, and the rifting keeps on chugging along, you may see oceanic crust appear further south.

The thin lithosphere of the Afar Depression, already weighed down by its magmatic plating, may keep on sagging. Water from the Red Sea will eventually tumble into it. The crust will split apart. Oceanic crust will be forged. In several tens of millions of years, East Africa will be no more. Africa will be a large but downsized continent. A new microcontinent, severed ever so leisurely from Africa, will now exist, left adrift in the Indian high seas. And between the two, born from the planet's most extravagant volcanism, will be the world's youngest ocean.

IV

THE VAULTS OF GLASS

When I was a kid, and I first found out that underwater volcanoes existed, I immediately imagined them as a perfect home for supervillains. Fine—as an apparent adult human, I still think of them in this way. Volcanoes on land are breathtaking enough, but the idea that you get volcanoes underwater is really quite sensational. Surely there is no better place for a nefarious mastermind than in a fortress literally made of magma. Not only are underwater volcanoes hard for pesky law enforcement, superheroes, or secret agents to find, but, being powered by the geothermal heat from below, incognito bases built inside them would be environmentally friendly. Plus, there's a lot of real estate, as most of Earth's volcanoes are underwater.

"There's more volcanic activity in the oceans than on land," says Bill Chadwick, a seafloor geologist at Oregon State University's Hatfield Marine Science Center. "That's not obvious at all, right? We pay attention to volcanoes erupting on land because they drop rocks on our heads. The ones in the oceans do so secretly, so you'd never know it."

There is a world of fire down there in the shadows. From exploding sunken mountains to lava-spewing cracks running through the seafloor, there is so much to see. That is far easier said than done, though.

An opaque ocean sits between us surface dwellers and that magmatic Arcadia. That's too bad, because if Earth's eruptions mostly happen in the deep sea, and we can't get to them, we aren't going to answer the most fundamental question about the planet; namely, says Ken Rubin, a volcanologist at the University of Hawai'i at Mānoa, "how does Earth work?"

IN THE NINETEENTH CENTURY, people tired of seeing maps marked with huge expanses filled with sketches of sea beasts. They wanted to know what was really out there. Some simply wanted to know how far down everything was. But how to find out? Some attached cannonballs or other lead weights to extremely long bits of rope, chucked them overboard, waited for a thunk, and took a measurement.

British oceanographer John Murray and Scottish natural historian Charles Wyville Thompson wanted to do a little better than that. After waving some sea creatures brought back from the Atlantic and the Mediterranean at members of the British government, they were funded to sail around the world and conduct science on the seas. They used HMS *Challenger*, a warship in the Royal Navy that had been converted into a floating scientific laboratory. From 1872 to 1876, from the Antarctic Circle to the Indian Ocean to New Zealand, they saw plenty, scooping up marine life and teasing out the chemistry of the seawater as they went.

They were also keen to find out what the seafloor looked like. Instead of a flat abyssal plain akin to a world-sized lake's bottom, they found that the seafloor had trippy topographic ups and downs. The greatest of the downs came courtesy of the Pacific Ocean's Mariana Trench, where a weighted cable kept descending for 36,000 feet before finding a floor. That spot, the Challenger Deep, is the deepest spot in the oceans, deeper than Mount Everest is high.

But the oddest underwater feature they found was in the middle of the Atlantic Ocean. In one spot, instead of a uniform landscape, they found that there was a huge wall[1] of some sort splitting the ocean into an eastern and western section. The dimensions of this wall couldn't be ascertained through the depth soundings they took, and it was left frustratingly unexplained.

This wall happened to be a mountain range. Those mountains would turn out to be volcanoes. And now we know the oceans are full of them.

YOU'VE LIKELY HEARD OF hydrothermal vents. In 1977, around the time NASA was sending robots to not just orbit but land on the surface of Mars, scientists were sending cameras down to the depths of the world's oceans, dragging them about and seeing what they could find. Near the Galápagos Islands, about 8,000 feet below the water, these cameras were nosing about along part of the seafloor that was cracking apart, occasionally erupting lines of lava. But, to the surprise of the 30 or so scientists that were watching from above the waves, the cameras also caught sight of a shimmering column of water emerging from a collection of rocks nearby.

Completely by accident, they had discovered the world's first known active hydrothermal vent field.[2] Two years later, and a couple of hundred miles north from these vents, scientists stumbled across the poster child for deep-sea exploration: black smokers, Dali-esque lithic chimneys erupting columns of dark, broiling liquid that gets its color from black iron sulfide. White smokers have been discovered too, which jettison fine particles of light-colored silicon, barium, and calcium.[3] And to date, hundreds of hydrothermal vent sites have been found all over the world, from top to bottom.

No matter the type, hydrothermal vents are all powered by the same mechanism: pockets of near-surface magma cook the surrounding rock, releasing all kinds of chemical compounds from it; the hot rock heats up percolating seawater, which gushes out of vents—holes, or chimneys—in the seafloor as a murky, 750-degree Fahrenheit fountain. Although way beyond boiling, the hot water remains a liquid because the weight of the sea pressing down on it stops it from gurgling and sizzling.

These famous spouts aside, most scientists don't know a whole lot about underwater volcanoes. That old adage, that we know more about outer space than we do about the deep oceans, holds true today. Having explored just a tiny percentage of the seafloor, the number of underwater volcanoes we think exist is nothing more than a guess. Depending on whom I've spoken to, there are tens of thousands of them in the world's oceans to as many as several hundreds of thousands. Many are extinct, while some are furious and seem clear on letting everyone know. Much like their terrestrial cousins, some erupt lava in a rather calm manner, which quickly takes on a silvery sheen as the cool seawater smothers out its embers. Others decide to take the explosive route, belching out gassy magma in one big rush that creates pumice, that light, holey rock you use to exfoliate.

Lava is often perfectly happy appearing around cracks in the earth, a bit like the fissures running through Kīlauea's rift zones. Sometimes, if the lava is squeezed out just right, these cracks can pump out little balloons of lava that snap off, quickly crust over, and tumble about on the seafloor. Geologists call them pillow lavas. I once rested my head on an old pillow lava. Don't expect to get a good night's sleep with one.

But the volcanic mountains themselves are far more impressive. Some of them are rather tall and pointy, while others are far wider than they are vertiginous. But in many visual respects, they aren't vastly different to volcanoes on land. Imagine getting a helicopter and flying around a volcano, like Mount Fuji. Now imagine that the world has flooded and instead of flying, you're swimming around Fuji. That is essentially what looking at an underwater volcano is like.

Scientists no longer drop cannonballs into the sea to find what bumps and troughs exist down there. Mostly, they use pulses of sound: they blast it from the base of boats or submarines at the seafloor, see how long it takes for the sound to get back to the boat, and use that to make topographic maps of the murky floor. What's that? A big hill on its own? You've probably found a submarine volcano.

But we're human. For us, seeing is believing.

AS WITH SO MANY volcano aficionados, the inspiration for Chadwick was the same: between his junior and senior years in college, Mount St. Helens exploded sideways and carved out a giant hole in Washington State. Being a geology major, he couldn't resist the pull: he got a bit of funding to study it and volunteered his time over Christmas break. "Before I knew it, I was flying in a helicopter into the crater and got the biggest jolt of adrenaline I'd ever gotten, and that was it."

Eventually, after completing grad school in California and postdoctoral work in the Galápagos, he found himself with a job at America's National Oceanic and Atmospheric Administration, which was looking to explore the seafloor but needed someone with a volcanology background. Chadwick was the perfect person for the job. At that time, in the late 1980s to early 1990s, seafloor volcanism was an unexplored niche. And so began decades of dives beneath the sea, with scientists such as Chadwick beginning to seriously look for underwater volcanoes and their damp eruptions.

When you think of scientists exploring the seafloor, you probably think of crewed submersibles, essentially see-through bubbles that slowly descend into the Cthulhuian depths. Chadwick himself made 10 dives in one such bubble affectionately named *Alvin*. "It was sort of like an astronaut going backwards," he says.

Like astronauts, they don't let any random person aboard. "What they do is they put you in the sub on deck, the day before, to see if you freak out or not," he chuckles. "They don't want that to happen on the seafloor!" The sphere is also just seven feet or so across and filled with electronics. "It's pretty tight in there," he says, so you can barely stretch out. Dark-humored pilots may not assuage the fears of the nervous. "There's enough air in there for three people for three days," he recalls one of them saying, before adding: "Or one person for nine days." Chadwick chuckles again. He tells me that there are reminders, both vocal and on a vanity license plate, that demand that you expunge your bodily waste before you go. There aren't toilets on *Alvin*, so anyone ambushed by the call of nature several miles underwater either has to hold it in for hours on end or has to take the undignified option and deposit their dregs in special bottles.

When everyone gets the thumbs up, these submersibles are launched onto the sea surface, their ballast tanks take on water, and you just sink down.

"I loved it, it was a thrill," says Chadwick. "My first dive, I remember looking out of the window, and seeing the seafloor for the first time was really, you know, mind-blowing, just the idea I was on the bottom of the ocean." He was exploring a part of the Pacific Ocean just off the coast of Washington and Oregon known as the Juan de Fuca Ridge, a site of underwater volcanism. There, it takes 75 minutes to get to the seafloor, and all the lights are left off to conserve battery power. The fields of glass, illuminated by *Alvin*'s lights, were certainly a sight to see. But like many stories, it's all about the journey, not the destination.

"You look out the window and the ocean is just full of bioluminescent, gelatinous organisms," he says. "As the sub is sinking down it's slightly disturbing them and so they're all glowing as you slink by them. As far as you can see, there's glowing things going by the window." He recalls sitting there, slack-jawed, transported straight into what appeared to be a sci-fi universe.

Unless you're looking for aquatic life, though, human submersibles aren't the main way scientists gape at underwater volcanoes. Remotely operated underwater vehicles, or ROVs—robots remotely controlled from the surface—are the go-to option because they can stay underwater for far longer to conduct more science and, well, there's no risk of a violent implosion event turning any humans into gory pancakes.

Richard Camilli, an associate scientist at the Woods Hole Oceanographic Institution, tells me about a meeting of scientists in 1986, just a few months after the Chernobyl nuclear disaster. The keynote speaker puts up a photograph not of a fish or an underwater volcano, but of a guy smoking a cigarette, baffling the audience. "Turns out he was one of the guys being sent in to shut down Chernobyl after the failure," Camilli says. "This guy knew it would be his last cigarette." The speaker used this single photograph to point out all the good robotic explorers could do in such a dangerous world. "It really transformed the robotic research community," he says. From that point on, ROVs were the name of the game. One major drawback, however, was their inability to stray far from the ship of humans that commanded them thanks to their electrical umbilical cords.

But in 1994, Woods Hole put together a robot named ABE, the Autonomous Benthic Explorer. ROVs looked at ABE with envy, for this podgy metal diver was unshackled from the chains of its human masters. Able to swim about literally untethered from its inventors, it had a rudimentary artificial intelligence that allowed it to follow a series of waypoints and use onboard scientific instruments to sniff around for geologic, hydrologic, or zoologic curios. It started out as a mobile seismometer, listening for quakes, but over the years it was able to detect magnetism, map out volcanic fissures and mountains, and even get up close and personal to some of the hottest and deepest hydrothermal vents. "They didn't have to look at the sensors, because it came back scorched," says Camilli. "It was in water, and it burned."

After clocking up 221 research dives, and with smarter autonomous underwater vehicles, or AUVs, nipping at its heels, ABE entered retirement. Or so it thought: with so much science to do, and with a research expedition off the shores of Chile looking like one hell of a send-off, scientists brought ABE out of retirement for one last job. And in 2010, while on its 222nd dive to the seafloor, it suddenly imploded. "It was so transformative in the field of robotics that the *New York Times* had an obituary for it,"[4] says Camilli. Such death notices are rare. The last time the *Times* gave a robot this honor was when a global dust storm on Mars killed off NASA's venerable *Opportunity* rover in 2019.[5]

ABE looks primitive by today's standards. "It was just made to follow the breadcrumbs, sort of like a wind-up toy," says Camilli. If you told him back when ABE was made the sort of things AUVs are doing today, he adds, you'd be placed in a padded room with a straitjacket.

He's right. The future is already here. In 2015, one named *Sentry*, under the watchful eye of Chadwick and company, was having a swim just south of the Mariana Trench, the deepest gash in the seafloor. Expecting to find old volcanoes, it instead came across a field of freshly made volcanic glass. It was the deepest underwater eruption ever found,[6] frozen in time, hiding three miles below sea level. Some of the glistening lava flows down there are nearly five miles long and 450 feet thick, higher than a 30-story building.

Camilli was part of a team that sent a small fleet of AUVs on a NASA-funded expedition to the Kolumbo seamount, a seriously angsty volcano just north of the Greek island of Santorini, itself a volcanic isle. In the past, Kolumbo has exploded so violently that it has killed people on Santorini with tsunamis and asphyxiating clouds of carbon dioxide. Today, this hydrothermal vent–adorned symbol of rage is a labyrinth of towers and spirals of frozen lava.

Camilli told me at the time that getting a robot to navigate through this hadean nightmare was a bit like hang-gliding in midtown Manhattan.[7] But that was the point: this expedition was designed to push these AUVs to their limits. If they crashed or tumbled into some lava, they would die.

Remarkably, all three AUVs—two smaller reconnaissance-type vehicles designed to scout ahead, and one larger more technically capable sub—survived the gauntlet. Their artificial intelligences not only allowed them to recognize danger and avoid it, but they also spoke to each other and shared information so they could carefully and efficiently plot out their scientific missions. I practically squealed with joy when Camilli told me the names of these robotic mission-planning systems. The boss AI was named "Kirk," while the one working out the scientific goals was dubbed "Spock."

Spock was perhaps a little too good at its job. The expedition's scientists, who could override the robots if need be, wanted them to check out a specific site, hoping to get some cool data. Asking Spock if this was a good idea, it shook its digital head and defiantly headed off to an unexplored sector instead. There, it found a brand-new site overrun with incredible hydrothermal events, a far more exciting adventure than the one the silly humans suggested. Scientists weren't even required to tell the robots to pick up samples. Using onboard cameras and its digital brain, the larger sub snatched tantalizing rocks all by itself with its own mechanical arm.

AUVs don't always make it back. It is, says Camilli, always nerve-wracking whenever you let these expensive machines go. "Sometimes we refer to it as the pucker factor," he says. "If they have a bad day, it's game over." At one point, thanks to some misunderstood sonar signals, one of the Kolumbo AUVs misinterpreted a column of lava as giant gas bubbles and smashed right into it, "like a slow-moving train wreck," he says. But it lived to science another day.

ROVs still outnumber AUVs, but times are quickly changing. AUVs can go places humans and ROVs cannot, stay there for weeks on end, and can operate faster and better than scientists piloting a robot ever could. Unsurprisingly, this technology is attracting military interest, but a large subculture within the scientific community prefers its robots to hunt for underwater eruptions, not people. Unfortunately, when it comes to spotting live underwater eruptions, humans and their robotic companions have done a terrible job. The oceans are too big for these little robots to stumble upon an actual ongoing eruption.

There have been plenty of close calls. Often scientists visit a site of known volcanism, see nothing, come back a few weeks later to do some more work, and find that the seafloor now has lava cooling on it. The products of that ultra-deep eruption seen by *Sentry* were laid down just a few months before the robot arrived. One site in the Pacific Ocean was visited so close to an eruption that the tube worms that once lived there were still smoldering when the scientists turned up. The site, says Chadwick, was nicknamed the Tube Worm Barbeque.

Researching an eruption after the fact is still helpful, but you can't hope to understand them without catching them in the act. But turning up, seeing the cooling lava, and thinking "oh crap, there's been an eruption!" is the default outcome, says Rubin, a talker as wonderfully effusive as the volcanoes he studies. That's regrettable, "because it is the most amazing thing to witness, when you can actually see the glowing hot lava, and the explosions." Being underwater, you can get incredibly close to the fires without harm, too. "It's spellbinding," he says.

IT'S A TESTAMENT TO the might, terror, and shock of the 1980 Mount St. Helens eruption that so many volcanologists I've spoken to cite it as their primary inspiration. Jackie Caplan-Auerbach, another of the volcano's acolytes, has infectious, unbridled enthusiasm for rocks that grumble and mountains that rumble.

Caplan-Auerbach studied astrophysics as an undergraduate but felt stupefied by the prep school members of her educational institution and presumed she simply couldn't do science. For a while, she taught physics at high school. She was also dating a fisherman back then. "I was really envious that he lived his life on the water," she tells me. "And I thought, well, I can't do science on the water, because the only science on the water is marine biology, and I hate biology."

One day, she went to a huge bookstore in Portland ("the greatest bookstore ever on the planet"), headed to the nautical books section, and found a book that said *Physical Oceanography*. "And I thought: holy shit," she says. "There's physics *of the ocean*."

On the same trip, she climbed Mount St. Helens with a friend. "You get to the top and there's just devastation. Awesome! And I said, this is it. I'm going to study earthquakes on undersea volcanoes and how they may or may not cause tsunamis. And I had no idea if there were earthquakes on undersea volcanoes. I had never taken an Earth Science class in my life." At 27 years old, she was starting anew. She cold-called graduate programs and eventually found someone who said he wanted to put a seismometer down on a local undersea volcano, the person who became her future advisor.

Today, Caplan-Auerbach is a seismologist and volcanologist at Western Washington University, and she *adores* underwater volcanoes like most people love dogs—the perfect person, I thought, to show me the rarest of things: an underwater eruption, caught on camera.

It's 2020, so we must have seen a fair few by now, right? Shockingly, scientists have only visually seen two—and maybe just one, depending on whom you ask. The first was caught on an ROV camera back in 2004 at a volcano named NW Rota-1,[8] 60 miles or so north of the Pacific island of Guam. Newly formed hot rock tumbled about as little explosions tore off chunks of the volcano, but no proper flows of lava could be seen.

But, says Rubin, a more impressive underwater eruption came courtesy of the West Mata seamount[9] near Samoa in the South Pacific. Its effervescence was first felt by various sensors in 2008, but in May

2009 an ROV named *Jason* saw it putting on a pyrotechnical performance like none other. Caplan-Auerbach gleefully shows me a video of it in action: huge bubbles of carbon dioxide wobble up from cracks of flame; to the right, what looks like a series of firecrackers is set ablaze before a rock seems to unzip, letting lava roar into the seawater before the crack suddenly seals up as if it never existed in the first place; elsewhere, booms rip rock apart as if dynamite were being used; showers of fresh lava, quenching quickly, rain onto the seafloor as beads of glass. The entire thing looks like it's alive, a creature reacting to a predator by rippling, twisting, and hissing fire.

Caplan-Auerbach has seen this footage countless times, but she reacts to the footage as if she had first seen it earlier that day. Sure, seeing eruptions after the event is fine. "But there's something so different about seeing it happen," she says. Rubin, who witnessed the West Mata eruption at the time it was discovered, says it was the pinnacle of his career.

We need to find more undersea eruptions. Not only are these volcanoes the true exhaust ports of the planet, the primary way it cools down, but they also provide a way to study eruptions—the sort that, on land, can kill people—safely. "Give how long we've been studying volcanoes, and how much we know about them, we are woefully ignorant about what these eruptions are like," says Caplan-Auerbach. "It speaks to how inaccessible the oceans are, because these eruptions happen all the time."

MARIE THARP, in her own words,[10] would never have had the chance to study geology if it wasn't for Pearl Harbor. As men were shipped off to war, women were needed to fill in the gaps left in society. When the geology department at the University of Michigan opened its doors to women and promised them future careers working in the petroleum industry, Tharp reckoned she had a shot. She helped hunt down oil in

Oklahoma for some time, but in 1948 she joined New York's Columbia University, where she was hired to help male graduates conduct their research.

By that point, Woods Hole had given the US Navy advanced echo-sounder technology, which allowed the Navy to continually bounce sound waves off the seafloor. When these sound waves returned to the ship, a stylus would burn a spot on a lengthy sheet of paper, indicating the depth at that point. As ships went back and forth, more and more depth dots were gathered. Tharp, like all women at the time, wasn't allowed on the research cruises herself, but she was asked to perform the seemingly monotonous task of literally joining up the dots and making continuous depth profiles of the seafloor.

After months of painstaking work on six depth profiles of the North Atlantic, Tharp painted a picture of a track of seemingly disjointed mountains right in the middle of the Atlantic Ocean, with a valley running through the middle of those mountains. But when she showed her work to a male colleague, he dismissed it, saying it looked too much like evidence for something called continental drift.

Decades earlier, German scientist Alfred Wegener looked at the planet and suspected something fishy was going on.[11] Fossils of tropical plants were turning up in the Arctic. The remains of long-dead reptiles found in Africa were also unearthed in South America, but there was no way they could have swum across the Atlantic Ocean. And the western shoreline of Africa sure looked like a mirror image of the eastern shores of South America. He proposed that the two continents were once joined up—and that, long ago, every continent was stitched together in a so-called supercontinent. But something had caused them to break up and drift apart, a theory that was referred to as continental drift.

Failing to come up with a mechanism that could explain why continents would move about, the scientific community dismissed the idea as claptrap. So when Tharp suggested that she had found a valley-incised mountainous ridge in the middle of the Atlantic, perhaps the site of a split between two continents, her Columbia colleagues thought she sounded like a heretic. In what has to be one of the most moronic uses of the insult, geologist Bruce Heezen told Tharp that her idea was "girl talk."

EVEN BEFORE NW Rota-1 and West Mata were caught setting fires, scientists knew there were volcanoes erupting throughout the oceans' abyssal depths. As well as sonar technology and ROVs being used to find out where the seafloor had suddenly changed shape thanks to a recent lava flow, underwater microphones named hydrophones could hear the gurgling of aquatic eruptions from many miles away.

But even if you suspect a distant submarine volcano is erupting, says Rubin, "you have to move heaven and earth" to get all the best experts and equipment to the right spot in the ocean as quickly as possible as soon as the wheels of scientific bureaucracy have allocated expeditionary funding. So far, scientists have had no luck, turning up after the fires have been thoroughly extinguished.

Sick of arriving late to the party, and tired of listening in from afar like rubbish spies, scientists eventually decided to set up an entire observatory on an underwater volcano. If you sail for about 300 miles west from the Oregonian coast, you'll be floating above it.

Down below sits Axial seamount. It's a huge volcano: its base could squash the entirety of Austin, Texas, while the walls of its horseshoe-shaped caldera could surmount the tall pillars of the Golden Gate Bridge. Its magma reservoir is almost as big as Manhattan. A highly active volcano, it has been caught erupting through an array of scientific equipment three times—in 1998, 2011, and 2015. Scientists of all kinds visit the volcano and study it—on ships using pneumatic air guns to bounce sound waves through the volcano to map out its magmatic pathways or with robot divers poking about its lava flows.[12]

In 2014, a huge network of cables packed with technological trickery was laid down on the seamount and stretched all the way back to the mainland so that scientists could monitor it 24/7. It is unquestionably the most comprehensively studied underwater volcano in the world.

And yet, despite the success of these technological tentacles, no eruption at Axial has been caught on camera. The 2015 one was missed because the eruption took place on a distant flank of the volcano that hadn't erupted in centuries, away from the observatory's eyes.

It may be some time before another live eruption is caught on film. For the most part, scientists will have to remain satisfied with capturing the ghostly shadows of eruptions. Thanks to seismometers, though, they are getting better than ever at knowing *when* hidden volcanoes blow their tops. Seismometers made to walk the plank and slip below the waves can hear the spooky tremors of magma cracking up the seafloor before erupting into the eternal night.

These mechanical marvels are preposterously good at eavesdropping. They can hear meteors exploding in the upper atmosphere on the other side of the world, the slightest of changes in the flow of rivers, and even the hum of humanity,[13] from the thundering clatter of traffic to the footsteps of children running to school. And back in November 2018, seismometers all over the world heard a rumble emanating from the Indian Ocean, between the eastern African continent and Madagascar.

Both Maya Wei-Haas—a brilliant writer for *National Geographic*—and myself heard it, albeit indirectly. On Twitter, seismologists often share funky-looking squiggles recorded by seismographs, some of which are intriguing, some of which are so-so. But back then, that Indian Ocean rumble caused the social media cohort of the world's seismologists to scream "What the hell was that?" in the most professional way possible. One jokingly blamed it on a sea monster. And that's when the two of us science journalists knew we had a good story on our hands.

We followed the story of the seismic signal—a deep growl peppered with higher-frequency blips—for months, writing multiple pieces for *National Geographic* and *Gizmodo* as geoscientists banded together and shared data freely online as part of an international sleuthing effort. It turned out that the rumble was coming offshore the France-administered island of Mayotte.[14,15] Eccentric signals detected by France's National Center for Scientific Research in the spring of 2018 had already alerted them to the existence of these submarine shenanigans, and in 2019 they sent scientists out to the site. Seismometers were dropped, and the seafloor was semi-continuously mapped out by boat.

The bestial bellowing heard around the world turned out to be a huge volume of magma destroying the seafloor through a volcanic eruption.

Maps of the seafloor in 2014 showed no volcano. But new maps made in the first half of 2019, around a year after the rumbles began, revealed that a volcano had grown out of the seafloor, one that was already 3 miles wide and 2,600 feet high. It was an astonishing sight, as if a giant ant hill had grown in the middle of your living room overnight. They hadn't seen it directly, but thanks to the nosiness of the world's seismometers, scientists had witnessed the birth of an underwater volcano for the first time.[16]

Robin Lacassin, a geologist at the Paris Institute of Earth Physics and one of the keen early listeners of Mayotte's seismic symphony, tells me that the volcano is made of so much lava that if you put it into a cube, it would be a mile high, twice the height of Dubai's Burj Khalifa skyscraper. Although slowing down these days, it was once erupting lava so quickly that it could fill up more than 1,000 bathtubs every single second. It is, to date, the largest known submarine eruption on record. "The eruptions at Axial are sizable, but they're not huge," says Chadwick. "The Mayotte one was gigantic. How does that happen?"

And yet this baby volcano's fury doesn't hold a candle to what was discovered hiding at the bottom of the Atlantic Ocean 70 years earlier.

SWEEPING ASIDE THE accusations of "girl talk," Tharp pressed on, gathering more data from better echo sounders as it came in, and continued to join up the dots. To make it sound like a simple if gargantuan puzzle does her work a great disservice. It wasn't a mere matter of connecting points on a blank sheet of paper. Huge gaps between data points needed to be filled in for these depth profiles to make any sense. Using her geologic education and a remarkable intuition verging on that of a Jedi, she made sketches of the seafloor's visual appearance on the basis of the available depth-profile data. She reproduced the actual ups, downs, and in-betweens of it rather than a series of jagged lines depicting sharp relief changes.

In the early 1950s, Heezen hired Howard Foster, a deaf Boston School of Fine Arts graduate, to map out the positions of earthquakes in the Atlantic Ocean. When he finished plotting them out, Heezen and Tharp saw that they formed a path going north to south in the middle of the Atlantic seafloor. Tharp's relief map of the Atlantic Ocean was overlain atop the quake map, and her mountainous ridge aligned perfectly with the earthquakes' epicenters. In the summer of 1953, Heezen conceded that Tharp was right: there was a huge valley-incised mountain range snaking through the middle of the Atlantic. Today, we know this continuously cleaved edifice as the Mid-Atlantic Ridge.

It is colossal. This mountain range stands nearly two miles above the seafloor, is 1,000 miles wide at points, and is 10,000 miles long, stretching from just south of the North Pole to the islets sitting above Antarctica. Aside from a few spots here and there, the majority of the Mid-Atlantic Ridge is underwater.

The East African Rift represents the beginning of a continent's destruction. The Mid-Atlantic Ridge is the end of that journey, the sunken wound left behind by the annihilation of an ancient supercontinent.

Robert Stern, a plate tectonics expert at the University of Texas at Dallas, tells me a story: sometime around 200 million years ago, give or take, all the world's continents were united as one big landmass, or supercontinent, named Pangea ("all Earth"). But rifts, like what we see today in East Africa, began to tear it apart. What is now North America tore itself from northwest Africa around 180 million years ago, followed 30 million years later by South America divorcing itself from North America. Soon after, South America broke away from Africa, unzipping from the south to the north. Around 60 million years ago, Europe and North America splintered apart. And what appeared in that huge gap between the continents, step by step? The Atlantic Ocean.

The divorce proceedings continue today. The Mid-Atlantic Ridge, the connected zippers that opened up the ocean, is still pushing the Americas away from Europe and Africa. Both halves are moving apart at around 0.8 inches per year. That doesn't sound like much on human timescales. But give it 50 million years and the Atlantic Ocean will be 620 miles wider.

And, just like along the East African Rift, the sundering of the seafloor is making a heck of a lot of volcanoes. As the tectonic plates on either side move apart, the hot and plasticky deeper mantle rises and gets extremely close to the surface. It decompresses, melts itself and the lithosphere above it, making a lot of magma. And that magma erupts along the Mid-Atlantic Ridge. In other words, this 10,000-mile-long range of mountains is essentially one continuous volcano, and its magma, cooling to make new slopes before being pulled east or west, becomes fresh oceanic crust. This is where the seafloor is made. It is how oceans are born, or at least where the bathtubs needed to contain them are forged.

This discovery alone was revolutionary enough, but Tharp didn't stop there. Plotting the locations of more undersea earthquakes, Tharp and company found that the Mid-Atlantic Ridge connects up with the Gulf of Aden and into the East African Rift. That's when it clicked: the Mid-Atlantic Ridge is just one section of a long line of volcanic mountains that join up like fractures on the surface of an eggshell. These rifts, sometimes in the middle of the ocean but sometimes squashed up to one side, make up a single 40,000-mile-long serpent of mostly submerged volcanoes that arc across the entire planet. They are strands on a web of wounds that mark where the world's crust is being ripped open, lines that flag the beginning of oceans and the ends of continents, the ripples made by a planetary-scale engine of creation and destruction.

The scientific community remained skeptical about the idea of a 40,000-mile-long chain of undersea mountains, but in 1957 Heezen gave a convincing talk at Princeton University unspooling the evidence. There, renowned geologist Harry Hess stood up and declared that Heezen (not Tharp, of course) had "shaken the foundations of geology."

But not everyone believed the rifts existed. One prominent holdout was famed oceanic explorer Jacques Cousteau. He dragged a camera attached to a sled across the seafloor of the Atlantic Ocean, presuming he would see nothing. But when it awkwardly bumped into a giant central spine of volcanic mountains before dipping into the rift valley, he was converted and ended up showing the footage to a captivated gathering of ocean explorers in New York in 1959. That same year, the Columbia University team's work, including Tharp's pioneering maps and sketches, was published by the Geological Society of America.

At this stage, no one knew much about the mechanics of the Mid-Atlantic ridge or understood how it got there in the first place. They had joined up the rift network around the world, and Heezen suspected that the rifts were made by the splitting apart of the crust. But the idea that this could cause entire continents to move around, create supercontinents and then break them apart, was still seen as outlandish.

Tensions rose. At one point, the director of Columbia University's Lamont Geological Observatory, who was vehemently opposed to the idea of continental drift, tried to force Heezen out and deny him access to the university's oceanic data. The ousting proved unsuccessful, but Tharp was fired. Fortunately, Heezen, who had built up a long list of international collaborators by that stage, continued to pay her through grants from the Navy, allowing her to continue her work from home.

It was now the 1960s, and revolution was in the air. *National Geographic*, inspired by an expedition across the Indian Ocean, wanted someone with artistic prowess to create a vivid painting of its seafloor. They hired Austrian artist Heinrich Berann after his young daughter wrote to them, boasting that her dad could paint better maps than those they had previously featured. Heezen and Tharp were hired to help make the painting scientifically accurate, and in 1967 the *National Geographic* panorama of the Indian Ocean's depths was published to great acclaim.

Bolstered by this remarkable act of science communication, one that sold the beauty and mystery of the seafloor to the general public better than all the other fickle, argumentative scientists ever could, the team then began work on a panorama of the entire planet's seafloor, one fed by a quarter of a century of high-precision mapping work. It was published in 1977, the same year robots were discovering hydrothermal vents for the first time—and just a few months after Heezen, while aboard a nuclear submarine during an expedition off the shores of Iceland, suffered a fatal heart attack.

The 1977 map is undoubtedly Tharp's (and the team's) magnum opus, her equivalent of van Gogh's *The Starry Night*. The reds, greens, and yellows of the continents promenade around rich, deep blues of the world's ocean basins, through which the elevated purple grooves of the volcanic ridge network anastomose. Both scientists and the public could, for the first time, see the world as a bunch of tectonic plates. For their work, Tharp and, posthumously, Heezen were awarded the Hubbard Medal, the National Geographic Society's top accolade, in 1978. Not content with ending her remarkable career on that insurmountable zenith, Tharp continued working until her death in 2006.

Susanne Buiter, a tectonics expert at RWTH Aachen University in Germany, co-organized a special session for an online conference in 2020 in honor of what would have been Tharp's 100th birthday. "If you realize what she drew the map from—I mean, it was just lines across the ocean," she says. "And she drew everything in-between. Scientifically it's already inspiring, but then you realize how she actually did this, and the time she worked in, the fact that she was a woman in a field with only men."

Despite her game-changing work, Tharp still wasn't allowed on research cruises until the late 1960s. She was never first author on any scientific paper. And in those days the fact that a woman was a scientific author at all was itself very unusual. Despite all that, "she didn't turn bitter or anything. She just did the science," says Buiter. "She just went for it."

FORTUNATELY, Tharp lived long enough to see her "girl talk" transform our understanding of the planet.

When oceanic crust is first made, and its magnetic minerals cool down below a specific temperature named the Curie temperature (which varies from mineral to mineral), the minerals lock in the directions of the north and south magnetic poles of the planet. That's helpful, because the planet's indecisive magnetic poles keep changing position. Our planet's liquid outer core, which sits below the mantle, contains churning lumps of iron and nickel that, by going up and down over and over, gives Earth its magnetic field. Over the course of hundreds of thousands of years, chaos in the outer core causes the north and south magnetic poles to flip, a bit like rotating a bar magnet 180 degrees.

This magnetic somersaulting became strikingly clear in 1962, when HMS *Owen* documented the magnetism of the Carlsberg Ridge, a section of the planetary mid-ocean ridge system located in the Indian Ocean. Scientists saw a zebra-stripe pattern extending away from both sides of the ridge, a pattern that was (almost) perfectly symmetrical.[17]

The rocks within some stripes showed magnetic north around the top of the world, while others showed magnetic north at the bottom of the world. These stripes supported the idea that the planet's magnetic field flips over rather abruptly on geologic timescales.

Those oceanic zebra-stripe patterns were subsequently found imprinted on undersea ridges all over the planet. It indicated that new oceanic crust was being churned out of these ridges over millions of years, with fresh crust, as it cooled, locking in the magnetic field direction as another stripe. This strongly bolstered the idea that the planet's surface was being ripped apart along a world-encircling network of ridges that continually manufacture the seafloor.

Around the same time, "magnetic compasses" found on land provided another key bit of information. Magnetic minerals in continental rocks from the same location, but of different ages, were pointing all over the place. These compasses seemed to indicate that the direction of magnetic north 100 million years ago was different than the direction of magnetic north 200 million years ago, or than today. But this wasn't because the magnetic north pole was wandering about, but because the continents themselves had itchy feet. When those older rocks formed, their compasses pointed north. The continents these rocks were part of then moved about, so by the time those younger rocks had formed, magnetic north was in a different direction. Scientists eventually realized that they could use this fact to track the migration of entire continents across the face of the world over billions of years.

This type of evidence, when combined with Wegener's ideas, Tharp's maps, and a panoply of additional diverse evidence from all kinds of scientists all over the world, gave rise to the theory of plate tectonics.

We now know that when tectonic plates, drifting atop the mushy parts of the mantle, split apart, you get oceanic crust through extensive eruptions taking place over hundreds of millions of years. The dense oceanic crust, meeting lighter continental crust, sinks into underwater trenches where it tumbles into the deep mantle and loses its captured seawater. That water rises above the doomed slab into a pocket of mantle. There, it allows some of the rocks in the mantle to melt more easily. That alchemy eventually makes the gloopy sort of magma that erupts onto the surface and builds explosive, pointy volcanoes, like Mount St. Helens. Although dangerous at the time, these eruptions make new rocks for us to live on. When two continents slam into each other, they crumple up and form mountains. And when the planet feels like cheating, it fires up a mantle plume, cooks the tectonic plate above, and makes huge volcanoes appear in the middle of the ocean or on land.

That is the 101 on how Earth gets made. And we wouldn't have gotten there without Tharp. Her legacy is being appreciated now more than ever,[18] both in terms of her scientific output and her perseverance.

THE WORLD'S HIDDEN MOUNTAIN range isn't a perpetual pyre. Lava isn't continually firing out of the middle of it, otherwise scientists would have seen it. The gradual forging of the oceanic crust and the stretching and spreading of the ridges means that the lava leaks out of the cracks, but just every now and then, not in perpetuity. So unless we keep a massive fleet of AUVs down there all the time, using some futuristic technology, much of the seafloor's volcanic fireflies will glow unnoticed. Scientists count themselves lucky, then, when a few sunken volcanoes manage to erupt so powerfully that they show up above the waves.

Iceland sits at the northern end of the Mid-Atlantic Ridge. Eastern Iceland is on the Eurasian plate, while Western Iceland is on the North American plate. If you ever go to Thingvellir National Park, dive into the clear, icy waters of the Silfra fissure. You'll be swimming inside the Mid-Atlantic Ridge, nestled between two tectonic plates that, at points, you can more or less touch at the same time.

The fact that Iceland is above water is odd, because rifting alone can't produce an island packed with volcanoes that rises miles above the seafloor. But Iceland's a cheat. Coincidentally, at that exact spot, a mantle plume is also blasting the base of the lithosphere, cooking up a whole bunch of magma itself. The combined forces of the plume and sundering rock has produced a true land of ice and fire: here, profuse eruptions of lava often mix with ice, sometimes creating dangerous flash floods, noxious gases, and long-lived plumes of steaming ash that can blanket the sky and get into plane engines and melt them. Where were you back in 2010 when Eyjafjallajökull, the most misspelled volcano of all time, engendered the most extensive shutdown of European airspace since the Second World War?

Volcanoes without this combination of a mantle plume and sundering rock—and even a few with it, like Axial—struggle to rise above the waves. But that doesn't stop them from trying. Most underwater lava flows are too dense to ever produce debris that will surmount the sea's surface, but volcanoes with trapped gas have a better shot at being noticed.

In August 2019, sailors near the South Pacific's Tongan archipelago were suddenly surrounded by a fleet of floating rocks a foot thick and covering an area of 40 square miles. A volcano deep below had explosively erupted, creating gas-filled, head-sized blobs of lava that quickly cooled into pumice in the presence of seawater. So little is known about the volcano responsible, though, that it is named, well, Unnamed.[19]

That outburst was nothing compared to the 2012 eruption of Havre, another underwater volcano just north of New Zealand's North Island. It produced a raft of pumice so massive that it could bury Washington, DC, twice over. It was first seen by a passenger on a commercial jet that was flying over the area, who quite rightly wondered why the Pacific Ocean had turned gray.

Scientists aren't exactly sure how underwater volcanoes erupt explosively. To get an explosion, the pressure of the bomb—our gassy magma trapped underground—has to be far higher than the pressure of the surrounding environment. On land, when you poke a hole in the bomb, the once-trapped gassy magma rushes out, its gases explosively decompress, and boom, there's your explosion. But the weight of the *entire freaking ocean* above underwater volcanoes is often so great that there isn't a huge pressure difference between the seafloor and the magma trapped below it. There will be no rush to freedom for this gassy magma. So how can Unnamed and Havre have blown things up all the way down there?

I checked in with Tobias Dürig, a volcanologist at the University of Iceland, to answer that question. He helped me out with my PhD, and we wrote a paper together. He's a cool guy who blows things up in a lab to work out how volcanoes explode in the presence of water.

He isn't the only one to do this. Other scientists melt some common basaltic volcanic rocks in a special oven, pour the liquid into a crucible, put a heat-resistant straw filled with water inside it, and stand back. "And then you shoot at it and it explodes," Dürig says, matter-of-factly. Forget flammable fuels. For volcanoes, water—the same stuff that stops you turning into a shriveled husk—is the most explosive substance of all.

After being poked into the crucible, the water in the straw is injected into the magma, and the water's outer layer quickly boils off. That steam forms a heat shield around the liquid water, keeping it liquid. You see this effect if you drop some water on an extremely hot cooking pan: the droplets of water don't immediately evaporate; instead, they dance across it for a few moments.

That bleb of steam-shielded water inside the magma would just fizzle away over time too if it was just left there. But magma isn't calm. It churns about. That's where the gun comes in: "shooting" the magma sends a shockwave through it, one that shatters the steamy heat shield, exposing the water bleb to the magma. And that water is now vulnerable. As it heats up within the blink of an eye, it decides to take down its magmatic foe with it.

The water angrily expands into its new space, blasting the magma out of the way, a bit like a hyper-powerful hydraulic pump. In microseconds, the magma is now a million tiny pieces, some no bigger than specks. The water then takes on enough heat from the magma that it violently vaporizes into steam. If you ever find yourself standing next to this sort of volcanic explosion, you will die either (a) because you will be extremely badly burned, (b) because a bit of magma shot right through you, (c) because the force of the explosion turned your organs into papier-mâché, or (d) all of the above.

Injecting water into magma can be tricky. In Havre's case, it seems like it got creative. Taking rocks made by the volcano, Dürig and company re-melted them a little in a lab to make them a bit like they were originally. They held them down, pooled some water above them, and blasted some pressurized gas under them to get a few cracks to appear. The water got sucked into the cracks and met the hot melty interior of the rock; the water got seriously cooked, violently expanded, creating more cracks for more water to go meet more hot rock, and voilà, boom, underwater explosion.[20]

But what makes the cracks appear in the primed-to-explode volcanic rock in the first place? "That's a good question," says Dürig—scientist code for "we don't know yet." Much work left to do, then, but the fact that this is something volcanologists actually do—that they can't be bothered to wait for a volcano to explode, so they make volcanic explosions themselves—is seriously cool.

Sometimes, a volcano manages to summon all its might and successfully pop above the sea. From 1963 to 1967, an eruption south of Iceland produced so much lava that it created an island named Surtsey, one that is now gradually being eroded by the waves and sinking beneath the icy gloom. In 2013, 600 miles south of Tokyo, a profuse eruption made a baby islet. It kept on pumping out lava, and within months it had grown so much that it fused with nearby Nishinoshima Island. And the entire Hawaiian Islands chain jumped out of the Pacific Ocean thanks to a mantle plume. Kīlauea isn't even the youngest volcano in the line of succession. That honor goes to the offshore Lōʻihi seamount, which will rise above the waves in the next 100,000 years or so, joining up with the Island of Hawaiʻi or forming the latest stand-alone jewel in the archipelago.

There is plenty we still don't know about the world's underwater volcanoes. But what is abundantly clear is that they are one of the reasons we have land to live on. Their eruptions, from those that took place billions of years ago to those happening right this very moment, provide us with volcanic fortresses that stand apart from the ocean. Without them, the world would be a little too blue and, consequently, far less habitable for us landlubbers.

The construction of these fortresses, though, can sometimes go horribly wrong.

ON AUGUST 26, 1883, Captain Watson was sailing on the British ship *Charles Bal* near an isle in the Sunda Strait, a body of water between the larger Indonesian islands of Java and Sumatra. A little after noon, huge clouds shot out of the isle's mountain. As reported in an 1884 issue of the *Atlantic*,[21] he heard a strange sound soon after, "as of a mighty crackling fire, or the discharge of heavy artillery," which repeated with increasing volume a little later. A "furious squall, of ashen hue," blasted from the mountain, and as "darkness spread over the sky," massive chunks of hot pumice hailed down over the ship, forcing the crew to dive for cover. As this gave way to a prolonged shower of tiny pumice, small enough to blind anyone looking up, the sky and sea became one single shadow, and the horizon vanished.

"The night was a fearful one," Watson recalled. He witnessed columns of fire dancing up into the sky from the isle's mountain before shooting back down to Earth; spheres of pale fire bounding about on the mountain's flanks; the wind so hot and sulfurous that it was choking the ship's crew; and embers falling on them like pieces of iron cinder. The sky looked like it was malfunctioning: "one second intensely black, the next a blaze of light," he said.

The next day, early in the morning of the 27th, the crew couldn't see a single sign of life on either side of the Sunda Strait. The apocalypse, they thought, had come. Except it hadn't. The worst was on its way.

Just before noon, a terrifying explosion rocked the isle, 30 miles behind them. A wave rushed from the isle and onto the nearby land, rising up dozens of feet as the water crashed onto the coast and rocketed upslope. Watching the devastation unfold, the sailors couldn't believe their luck: the wave coming from the exploding isle passed under their ship without being felt, seemingly sparing them from death. But the terrors kept on coming. Night appeared in the middle of the day as mud and sand-like matter cascaded down onto the ship. It was so dark that the crew could no longer see each other.

"Such darkness and such a time in general, few would conceive, and many, I dare say, would disbelieve," Watson noted. By the morning of August 28, the shadows began to recede, "and then was seen the result of this paroxysm of nature," he said. "The northwestern part of Krakatoa Island had disappeared."

Watson was one of several witnesses to the cataclysmic eruption of Krakatau (misspelled at the time, and forevermore, as Krakatoa). Mount St. Helens's 1980 eruption was famous because the blast came out of, and sheared off, an entire flank of the volcano. But in August 1883, Krakatau blew itself up: that huge explosion, the apex of a multi-month eruption sequence, obliterated two-thirds of the entire island.

The total energy of the eruption's major blasts has been estimated to have been equivalent to 200 million tons of TNT. How powerful is that? On August 4, 2020, highly volatile explosive material left criminally unsecure in Beirut's port exploded, killing hundreds of people, injuring thousands, and reducing a great deal of the Lebanese city to rubble. The energy release of those explosives was estimated to be equivalent to 400 tons of TNT.[22] Krakatau's 1883 self-destruction was a jaw-dropping half a million times more powerful.

The sound of the biggest explosion of the eruption was so loud and energetic that it circled the world several times and was heard as far as 3,000 miles away from the volcano.[23] That's like being in San Francisco and hearing an explosion in Hawai'i, or being in Iceland and hearing the crack of a blast in North Africa.

The wave that passed effortlessly by the *Charles Bal* was one of several gigantic tsunamis, lateral movements of water that only truly reveal themselves when they pile up on shorelines. Largely because of those tsunamis, 36,000 people in the region died before summer's end.[24] And all that was left of Krakatau was the rocky wreckage of itself, volcanic shards sticking up above the water like shipwrecks.

Krakatau had destroyed itself. But the magma that made it lived on. Like a volcanic phoenix, eruptions at the site continued until a baby volcano rose above the waves in 1927. It was named Anak Krakatau, meaning "Child of Krakatau." And for nearly a century, it continued to grow out of the water.

History has a habit of repeating itself, and volcanoes are no exception. After a lengthy period of minor eruptive coughs, Anak Krakatau exploded on December 22, 2018, generating another tsunami. The waves killed 400 people on Java and Sumatra while 47,000 people lost their homes.[25]

A team of scientists, led by volcanologist Rebecca Williams of the University of Hull in England, used satellite imagery and models to piece together[26] what happened that December day. It seems that after a century of wonky construction work, the lopsided volcano had become structurally unsound. Its western flank collapsed into the sea, generating a tsunami and opening a new undersea volcanic vent. The magma and water mixing caused huge explosions that cleaved off the summit of the volcano, littering the sea with debris the size of office blocks.

The volcano, as it has done seemingly forever, is once again rebuilding itself through eruptions, rising above the Sunda Strait. And so the cycle continues.

BRENNAN PHILLIPS, who designs instruments for ocean exploration at the University of Rhode Island, had a tough act to follow. While going through grad school, Phillips was an ROV engineer for Robert Ballard, who can't seem to stop discovering secrets on the seafloor: he helped find the wreck of RMS *Titanic* in 1985, the gigantic Nazi battleship *Bismarck* in 1989, and was even part of the crew that first discovered the existence of hydrothermal vents back in 1977.

A few years back, Phillips was part of a team working in the seas around the Solomon Islands in the South Pacific. A submarine volcano named Kavachi was nearby, one known for its turbulent eruptions. "And I think to myself, there must be some cool stuff around there to check out and explore," he says. The paucity of the scientific literature on the volcano, he thought, meant that they were near something worth exploring. He asked the ship's captain to head on over, but the unpredictable violence of the volcano made them too skittish to dare. "We never got within five kilometers of the damn thing," he recalls.

Fortunately, someone on the ship knew of a guide in the Solomon Islands, an Australian named Corey Howell who goes to the volcano on nothing more than his personal boat. He tended to visit Kavachi once a year, sometimes to fish, and he was certainly happy to take some curious scientists to this sizzling seamount. Phillips secured funding from *National Geographic* to head back to Kavachi in 2015 on Howell's teeny boat, sail over the crater—the barrel of the volcanic gun that could go off at any moment—and essentially throw a bunch of scientific instruments down inside the volcano, including a camera that would be retrieved later as it floated back to the surface. Kavachi, which spends most of its days erupting, was having a nap back then, but they knew it could wake up and get going again at the drop of a hat.

"It was a like a two-minute job," Phillips says. "Let's get the boat over there, chuck the camera in, and get the heck out."

They managed to complete this death-defying mission, not once but twice. The first time around the camera bounced off the crater rim and didn't capture anything interesting. But, like Luke Skywalker's proton torpedoes making it into the tiny exhaust port on the Death Star, their second attack run was a success: the camera fell 150 feet into the crater, through increasingly hot and acidic water, and stayed down there. An hour or so later, it popped up above the water, and when it floated away to a safe distance the crew scooped it up.

They didn't quite know what they expected to find in the hellish volcano's maw when they looked at the camera's footage. But no one expected to see sharks, from the silky shark variety to scalloped hammerheads. They expected to see life outside Kavachi. "We know it's a sort of oasis," says Phillips. "But to see them inside there is totally different, not expected . . . not right, based on what we understand should be there."

When the volcano next erupts, any sharks caught inside will be incinerated, so it isn't clear why they are risking it during its short-lived spells of silence. Phillips hopes to get some tracking sensors attached to these sharks in the future, so their journeys and behavior can be compared with the volcano's activity. But for now, no one knows why the sharks like swimming about inside Kavachi.

In any event, it's clear that sharks and similar beasts, whose evolutionary lineages stretch back hundreds of millions of years, really seem to love underwater volcanoes. Not too long ago, Phillips was part of an expedition that sent ROVs poking about around the Galápagos. Completely by chance, they spotted more than 150 egg-cases[27] in and around a hydrothermal vent field. The largest collection of eggs was found being bathed in murky water venting from a black smoker nearby.

The fossil record shows that certain dinosaurs used to lay eggs near hydrothermally heated spots of soil. Today, some large birds—the descendants of certain dinosaurs—like the Tongan megapode burrow their nests in soil that is being volcanically baked. Like birds that sit on their eggs to keep them warm, it is thought that this sort of behavior helps the embryos inside the eggs grow properly, and quickly, into healthy babies.

The eggs found in the Galápagos's waters turned out to be those of a type of deep-sea skate, and the colors of the casings seemed to show the eggs at different stages of development. The team can't say for sure, but this nursery, the first of its kind found inside a hydrothermal vent field, is likely there to accelerate the embryonic development of the skates' eggs.

It's not just ancient families of fish that love underwater volcanoes. Seamounts, both active and extinct, are one of the greatest places for life of all kinds to make themselves at home. Lissette Victorero, a marine macroecologist at the Norwegian Institute for Water Research, tells me that they are riddled with life, from sponges, gardens of coral, shrimps, sea stars, brittle stars, lobsters, and, of course, lots of fish. There is always a high concentration of foodstuff drifting around seamounts, and the hard rocks of the volcanoes provide anchor spots for less mobile life-forms that feed off things flitting by in the water. Seamounts also have long slopes that go from shallow waters to deeper waters, which gives pickier critters a wide range of accommodation choices.

"A lot of people call seamounts biodiversity hot spots," says Victorero. "But because we don't know that much about the deep sea—we've sampled 0.001 percent of the deep sea—it's hard to say if they are true hot spots." But, she says, it's sure looking that way. Seamounts are jungles, teeming with so much life that a new creature is discovered every time ecologists take a tour of one.

Even the dangers of an eruption don't stop life taking hold. The Kolumbo volcano, the one Camilli sent Spock and Kirk through, produces a lot of carbon dioxide, which binds with water to become an acid. "Some parts of Kolumbo may be as acidic as Coca-Cola," he says. "A fish swimming around there would have its bones dissolved." But what he and his colleagues are seeing around death-dealing sites like this are punctuated regime changes. When a volcano erupts or belches out a lot of carbon dioxide, it evicts a lot of life, allowing more resilient creatures to move in. When the volcano calms down again, the situation is reversed. "It's almost like seasons, but they're happening on maybe a weeks to months kind of scale, maybe even shorter," he says.

Seamounts also act as waystations for animals making voyages across the ocean. Sea turtles coming from the Great Barrier Reef use Kavachi, of all places, as a transit point to fuel up before moving on. And a type of shrimp first identified on Lōʻihi was later found on West Mata, meaning it somehow finds its way thousands of miles through the ocean with no other volcanoes in-between. Undersea volcanoes can even protect life against climate change. Shallow depths are warming quicker and acidifying faster as we continue pumping greenhouse gases into the sky. Seamounts deeper down will likely act as castles for biodiversity, says Victorero, as the rest of the planet burns up.

But these fortresses may not be sturdy enough to stop certain humans breaching their walls. Seamounts contain valuable minerals used to power the world's technology, and mining companies are already looking to exploit that. "There's about 200 seamounts out of a potential 34,000 that have been physically sampled and have had biological observations," says Victorero. If mining begins, we won't even know what they're destroying. What if we are destroying an ecosystem that's not present elsewhere? What waystations are we taking off the map?

The technology required for deep-sea mining is still being developed, but it won't be long before seamounts start getting dismantled. And the deep-sea environment is so far from our sight that it could easily be exploited by well-funded agents. "Industry can do a lot of damage without us scientists ever finding out. And a lot of these places are in areas beyond national jurisdiction. There's no law there. It's really hard to pinpoint who has the right to do what and how they're supposed to do it," says Victorero. "Once you open the door to do that, I would be very worried. Well, I *am* very worried."

ARGUABLY, the most important type of life enjoying underwater volcanoes is the stuff we can't actually see. The discovery of hydrothermal vents in 1977 was remarkable, but it was what was found living inside them that really set off an ecological revolution: millions of microbes.

Microbes that can thrive in this seriously hot, high-pressure realm get their energy from the volcanic matter itself: volcanic gases, the minerals in the volcanic rocks, and the myriad of particles ejected by hydrothermal vents. They latch on to whatever is available—hydrogen, methane, iron, sulfur, carbon dioxide, whatever. And they use this matter either as food or as ingredients they can use to cook up chemicals that become their food.

On the surface, and in much of the ocean, the base of food chains is plants or bacteria that can slurp up sunlight and use it to make energy—in other words, photosynthesis. But deep down, life has fully embraced the darkness. These microbes can exist at temperatures that would normally boil water, but most appear to be comfortable in hot, but not too hot, hydrothermal vents fields. And there, microbes form the base of undersea food chains, which support tiny animals, which in turn provide a food source for the more charismatic bigger beasts of the deep. "Animal life would probably not exist around a hydrothermal vent without a microbe that has the ability to capture that volcanic energy, to fix their own carbon dioxide, and to build carbon that then supports an ecosystem," says Julie Huber, a marine microbiologist at the Woods Hole Oceanographic Institution.

"One of the most exciting discoveries of the century was finding chemosynthetic life at submarine volcanoes—a whole new kind of life," says Chadwick. And it meant that ideas as to how life got started on Earth in the first place needed a rewrite.

There are all kinds of places that could have given rise to simple microbes billions of years ago. "There is a huge array of possibilities. We'll probably never know, and that's okay," says Huber. But the first microbes, from which everything else has descended, were not photosynthetic. They didn't drink sunlight. They breathed volcanic fire. They were probably a little like the hydrogen-eating microbes that are found all over the seafloor today.

Underwater volcanoes are not natural weapons of mass destruction. "They are effectively incubators," says Camilli. No matter the cause—from prolonged, epic, continental-scale eruptions to asteroid impacts—the mass extinction of life is largely driven by a profoundly perturbed climate. But seamounts and hydrothermal vents always seem to provide a Goldilocks region where conditions remain just right for microbes to survive. "Their volcanism outlives the cycles of climate change," he says. Life will prevail around the cauldrons of the deep.

And, as scientists have only recently discovered, life will also find homes *inside* these cauldrons. Every single expedition over the past decade that drilled deep into oceanic crust, both freshly forged and staggeringly ancient, found these volcanic rocks to be riddled with entire civilizations of microbes either thriving, surviving,[28] or existing in a state of suspended animation,[29] depending on the availability of nutrients in their surroundings. These microbes aren't like animals. They don't exist on human timescales. They persist on geologic timescales. They can hang out in the most forsaken parts of the planet's underworld. And it's starting to look like the oceanic crust is one of the largest habitats for life on Earth.

"When I was younger, I think we felt that life was really tenuous, we thought it was kind of a miracle," says Caplan-Auerbach. "Now what we know is that life exists anywhere it can find a niche. There will be something that says, oh, no one's living there? That's going to be mine."

Undersea volcanism, then, may have inadvertently incubated life while providing a habitat for it that is protected from the very worst catastrophes. It's strange, then, that despite their importance and their flair for the dramatic, submarine volcanoes haven't found their way into the zeitgeist. "The public is more interested in space exploration than deep-sea exploration," says Victorero. Everyone knows what NASA is—it's a brand as much as it is a space agency at this point—and the public's general fondness for space missions permits NASA and its sister organizations around the world to explore the night sky. One can only imagine what could be discovered if underwater volcanoes got that sort of hype. For now, "it's hard to get funding to purely go and explore what is on our own planet," she says.

But space science has taken a cue from those sunken crucibles. The discovery of these volcanic microbes, says Chadwick, hasn't just transformed how we look for life on Earth, but how we search for it elsewhere in the universe. If life doesn't need sunlight to survive and can live in a volcanic twilight at the bottom of the oceans, then perhaps planets that appear to be deadly to animals and plants may still be perfect places for microbes. As drilling into frigid oceanic crust demonstrates, you might not even need hydrothermal vents or active seamounts to provide a home for them.

Kennda Lynch, an astrobiologist and geomicrobiologist at the Lunar and Planetary Institute in Houston, Texas, is one of several scientists searching Earth for environments that may be similar to those we find on other worlds. And, she explains, that cramped environment beneath the seafloor can be found elsewhere: on a planet with a crust made of the same type of volcanic rock, a world once covered in erupting volcanoes, including those underwater. This planet's flowing rivers and active volcanism mysteriously vanished during an Armageddon-like event. Its underworld, though, may still be warm and waterlogged. Until we dig deep into the crust, no one will know for sure. But if there was ever life on Mars, it may still exist below the surface today, hiding in the same sorts of volcanic tombs we find on Earth.

It may sound like a longshot. But our Atlantean volcanoes have revealed that once it gets going, no matter what you throw at it, life lives on, dreaming in darkness within vaults of glass.

38

V

THE PALE GUARDIAN

It was cold, as it had been for billions of years. Without wind to move them, the silver seas remained calm. Starlight glistened high above. The ground creaked like an aging wooden ship in motion, but the mountains remained still. The chasms remained quiet. Coliseums of rock stood in memory to eons of pandemonium past. Every so often, a shard of ice or stone from the edges of existence would rain down onto the surface and shatter upon impact. Stone snow would rocket up and float across the land, settling in silence somewhere else on this magnificent desolation.

But one day, a visitor fell from the stars. This one wasn't made of rock or ice, but metal. It was a ship, sailing from another world, decorated in the crimson flags of an alien nation. As it approached, a plume of orange rocketed out, scattering sunlight against the shadows. Showing no signs of slowing down, the vessel careened toward the Sea of Serenity, before slamming into the Marsh of Decay.

Ten years later, another wayfaring vessel sailed toward the argent realm. This one was different: it slowed down dramatically as it found a safe spot to drop anchor. After skipping over a gaping maw flooded by darkness, it found a spot in the Sea of Tranquility. Rocky flecks shot out from beneath as it set itself down.

Silence. Hours passed. Then, a hatch opened, and two boots poked out of it. Someone climbed down the ladder, paused, then placed a foot into the sea. Another pause, then the traveler spoke: "That's one small step for man, one giant leap for mankind."

Ten years after the Soviet Union intentionally crash landed its uncrewed spacecraft, *Luna 2*, onto the Moon, the United States managed to deliver humanity itself to the lunar surface. After 4.5 billion years of protracted isolation, this lonely outpost in the starry ocean finally had visitors.

The inspirational, soul-shaking power of *Apollo 11* is arguably no less potent today than it was half a century ago. The Apollo program's results were nothing short of magic and have firmly etched themselves into the imaginations of millions who were born long after they transpired. But what gets far less attention—at least, when it's not part of an eclipse—is the Moon itself. To many, it's done-and-dusted: we've been there, grabbed some rocks. It's ours now. And unlike the other worlds in our galactic neighborhood, it doesn't seem to *do* much.

That last part is somewhat true. The Moon is, today, a cold desert. It doesn't contain alien life. We've checked. Mars, the first planet humans are set to walk on other than our own, appears to be a more dramatic place perhaps more worthy of our attention. But look a little closer, and you'll find that the Moon really isn't what you think it is.

In March 2020, the US Geological Survey published a comprehensive geologic map of the Moon[1]—an amalgamation and streamlining of six preexisting maps released in 2013—the chief purpose of which is to provide a clear summary of what we know about the rocky sphere's geologic makeup. On it, you can see the Sea of Tranquility, or Mare Tranquillitatis, painted in splotches of purple and red. These hues correspond to basalt, a volcanic rock formed out of the cooling of lava, akin to what Kīlauea churned out so eagerly in 2018. That means Neil Armstrong and Buzz Aldrin, the first ever human emissaries to another world, did, in fact, land in a genuine sea, as Michael Collins looked on them from above. It just happened to be one made of frozen lava, stretching 540 miles from one shore to the other.

Zoom out, and that geologic map turns into a multicolored marvel of volcanic deposits. The nearside of the Moon, which always faces Earth, is covered in similar frozen volcanic seas, or *maria*. Ancient lava flows can be seen scattered across the lunar surface like confetti. Elsewhere, splashes of oranges, reds, pinks, and purples reveal fields of volcanic glass.

"The Moon has been geologically almost dead for a long time," says Sara Russell, a planetary scientist at London's Natural History Museum. But, she is quick to note, it is also a volcanic mausoleum, a collection of volcanic mountains, cones, domes, valleys, rivers, channels, tunnels, and craters that owe their existence to a global ocean of magma.

Our pale guardian is more than just a frozen eruptive wonderland in space. Earth's earliest memories, lost from the planet long ago, remain etched in lunar stone. If we can get our hands on some of it, we will be able to open a window, billions of years back in time, to the very first days of the only home we've ever known.

SOME TIME AGO, out in the wilds of Antarctica, Ryan Zeigler found out that NASA was advertising for a one-of-a-kind job: the curator of the Apollo lunar samples at the Johnson Space Center in Houston, Texas. He also found out that he had six hours left to apply, so he rushed back from the field to McMurdo Station, an American scientific research outpost, and hastily wrote up his CV and application from scratch "in a crappy computer room."

He got the job. Today, he is not just the Apollo Sample Curator, but the manager of the Astromaterials Acquisition and Curation Office, which means that he also oversees NASA's entire collection of space stuff. That includes material from other planets, asteroids, comets, and even stars. In other words, he is the curator of extraterrestrial paraphernalia, which, if it isn't already, should proudly be emblazoned on his business cards.

The Apollo samples were thieved from the Moon during Apollo missions 11, 12, and 14 through 17, which were carried out between

1969 and 1972. They were collected at the various landing sites by astronauts, many of whom were trained by geologists to know what to look for. One astronaut, Harrison Schmitt of *Apollo 17*, was a professional geologist. In total, the astronauts brought 842 pounds of lunar material back home. (And not all of the Apollo samples turned out to be lunar material: a rock picked up by *Apollo 14* turned out to be a piece of Earth,[2] jettisoned into space eons ago before falling onto the Moon as a terrestrial meteorite.)

Most of the Apollo samples are stored in a hyper-secure vault. Fifteen percent of the samples are housed elsewhere in a secret off-site facility as a redundancy measure. That way, if a disaster befalls Johnson Space Center—a terrorist attack or a major hurricane, or someone gets inspirationally clumsy, for example—or if we find out that the samples have been contaminated in some unexpected manner, scientists still have pieces of the Moon to study.

These stones are some of the most valuable materials on Earth. As you might expect, you can't simply wander in and lick them. A rigorous, multistep decontamination process takes you, adorned in plastic shoes, a nylon suit, and a hair-ensconcing hat, through rooms with increasingly filtered air. When you get into the vault itself, you are greeted with the sight of multiple transparent cabinets. The Apollo rocks are kept in sealed containers, themselves inside sealed bags, in an atmosphere of nonreactive nitrogen gas to prevent the humid Houston air from reacting with the iron content of the rocks. They are snug and safely kept from dirty human hands.

As overseer, Zeigler's job isn't to smash rocks apart with a mallet and find out what the Moon is made of. That is the job of specialists within and outside of NASA, scientists that must demonstrate beyond all doubt that they deserve access to less than a tenth of an ounce of lunar rock in order to advance our understanding of the universe. Zeigler, who must always keep abreast of the most current lunar science, spends a lot of his time talking to scientists, helping them find the right samples while keeping the collection preserved for future generations. "There's more bureaucracy than I would like," he says. "But it's a government job, so what can you do."

The Apollo samples, and the chemistries they imprison, are our main source of lunar knowledge. The Soviet Union also pulled off three successful robotic missions to the Moon that snatched up lunar matter and brought it back to the motherland during the 1970s, but these samples were tiny compared to the efforts of Apollo. We stand on the shoulders of Apollo.

We haven't always had to go to the Moon to get shards of it to study. Sometimes, the Moon comes to us in the form of meteorites. Small space rocks, unimpeded by the Moon's lack of atmosphere, regularly slam into the surface[3] at tens of thousands of miles per hour, flinging Moon debris out into the night. Some of that debris lands on Earth.

Many lunar meteorites have been found in Antarctica, a continent whose vast, unpopulated, white expanse permits black, shiny rocks from space to clearly show up at the surface. In fact, the very first recognized lunar meteorite was found down there in 1982. By 1996, we had found 12. Katherine Joy, a meteorite hunter at the University of Manchester in England, says we now have more lunar samples in the form of meteorites than we do in the shape of Apollo samples. But it's not always easy to tell exactly where on the Moon they came from. They don't exactly turn up with a return address stamped on the front. That means meteorites are often like quotes: taken out of context, presenting the viewer with a skewed perspective.

With little lunar matter to go around, scientists have also made synthetic Moon rocks to play about with. Tim Grove, a planetary scientist at the Massachusetts Institute of Technology, was one of the very first scientists to study the Apollo samples back in the early 1970s. But it wasn't long before he and his colleagues used their knowledge of the Moon's chemistry gained from these samples to cook up their own batches of lunar lasagna. "You can make a synthetic Moon rock really easily," he nonchalantly tells me. You play about with different compounds—a pinch of iron oxide here, a cup of titanium oxide there—and eventually, you get a rock that can replicate not just the chemistry, but also the textures and crystalline structures you see in the real deal.

This sounds like witchcraft to me. "It's an art," Grove replies, with a slight shrug.

Telescopes on Earth keep an eye on the Moon too, as do a host of missions—some now dead, some still alive, some upcoming—that orbit above the lunar surface and use a suite of mechanical eyes to study the rocks below. NASA is particularly skilled at catapulting lunar observatories at the Moon.

Its *Gravity Recovery and Interior Laboratory* (*GRAIL*) mission sent two spacecraft, *Ebb* and *Flow*, dancing around the Moon a few years ago. A complicated mixture of geology, the Moon has some denser bits that have a slightly higher gravitational pull than other less dense Moon rocks. By detecting these tiny gravitational variations, *GRAIL* was able to create a 3D map of the upper slice of the Moon's underbelly, permitting scientists to see below the surface.

Like synthetic Moon rocks, *GRAIL*, to me, is scientific sorcery. But my favorite observatory has to be the Inspector Gadget–like *Lunar Reconnaissance Orbiter* (*LRO*). Launched in 2009 and still going strong, the NASA spacecraft is equipped with a tranche of technical trinkets. Some use laser beams to build up relief maps of the mountains, craters, valleys, and rings of debris; some are able to take the temperature of the lunar surface; some can detect subatomic particles sneaking off the soil. But I love it for purely aesthetic reasons. If you have ever seen a beautiful shot of the Moon's tapestry of shimmering milky devastation on the news or online, it likely was snapped by *LRO*. Short of abusing certain recreational drugs, *LRO*'s photography is the closest we will come to bouncing about on the Moon itself.

Many of the Apollo samples are volcanic. Not coincidentally, *LRO* has built up a breathtaking catalog of primordial volcanic features on the Moon, from explosion craters to static flows of lava. Julie Stopar, who co-runs the LRO Camera instrument, hadn't paid the lunar surface much attention in the past; she tells me she only had eyes for Mars once upon a time. But as soon as she peered at the Moon from the heavens, it clicked. "The Moon is a lot cooler than most people think," she tells me. "I don't think people know there is a lot of volcanism there."

As *LRO* helped make clear, much of the Moon's ancient lava has been made to look like Swiss cheese by billions of years of meteorite impacts. Scientists think they know roughly how many rocks punched holes into it over the past few billion years, particularly because there is nothing to erode away those old craters. That means a surface with more holes in it is older than another with fewer holes. This method, known as crater counting, allows scientists to estimate the ages of rocks they will never touch.

More precise ages are obtained from the Apollo and meteorite samples themselves, which contain radioactive elements. Over time, they shed particles, transforming or "decaying" into different elements, releasing heat as they do so. Each element decays differently, with each step of the decay happening over set, known periods of time. Scientists can look at an element, count how many decay steps it has made from its original element, and work out how long the entire decay has been going on for. The radioactive elements in rocks, then, are stopwatches.

All of this information is plugged into computer simulations and mathematical models to try and figure out what's what. But our comprehension of the Moon is still firmly built on the foundations of those Apollo lunar samples, and unfortunately, those samples are a bit rubbish.

MARIE HENDERSON HAS always loved the Apollo missions, right back to the time when she first found out about them as a kid. She carried that enthusiasm all the way through her life and into her doctoral thesis, something she was on the verge of defending at Purdue University when we spoke. Her PhD has been carefully tailored: she has chosen interesting sites on the Moon to study that would also make good candidates for visitation by future astronauts. Ideally, she'd be one of the moonwalkers herself. You'd think that she would hold the Apollo landing sites in high regard, particularly as most of them were inside frozen lava seas like *Apollo 11*'s Mare Tranquillitatis.

You'd be wrong.

"It's not that they were *that* boring," she says. "But they were, essentially, boring."

The landing sites were partly chosen because they were flat, a useful feature of many of the maria. Most of them contained a pretty similar rock, basalt, all taken from a relatively small swath of the nearside of the Moon. This means not only do we have a limited number of samples, but also they aren't even that different.

Imagine you were an alien geologist visiting Earth to collect some rocks to understand the entire planet. Instead of taking samples from all over the place, you only went to the flat unbroken plains of the Australian Outback, the American Midwest, and Siberia. Upon your return, your alien geologist boss might not be so pleased. There is no way you would be able to determine anything significant about the planet's multibillion-year-long history that way. The same applies to our endeavors on the Moon. Or, as Joshua Snape, a planetary geologist at VU Amsterdam in the Netherlands, puts it: "As a terrestrial geologist, you'd freak out if you saw the size of the samples we look at."

No matter how complicated and serious sounding your ideas about the Moon may be, they have their limits. "Models are great, and they tell us things samples can't," says Zeigler. "But without samples to provide ground truths to the models, the answer is whatever you want the answer to be."

Compared to Earth-based science, planetary science is a different beast. "You have to be ready to have some vagueness on your edges and use your imagination in a way to fill some gaps," says Henderson.

And because nothing in life is fair, the largest gap in our understanding of the Moon and its ancient volcanoes appears right at the beginning of the story.

YOU CAN WRITE entire books about the possible origin stories of the Moon. No one can say for sure where it comes from, which is a fundamentally strange thing to say. There are a few competing ideas: Earth and the Moon forming simultaneously out of a cloud of vapor, for example, or a long train of meteoroids smashing bits off our planet that would go on to make the Moon.[4] For the sake of simplicity, we'll start with the idea that is the least complicated. This increases its likelihood of being correct, for the universe tends to dislike needlessly complicated stories. It also helps explain a key feature of the lunar samples: lunar and Earth rocks have very similar chemical fingerprints, suggesting they are largely made of the same stuff.

The most straightforward way to explain this is that, 4.5 billion years ago, not long after Earth had formed from space rocks clumping together under their own immense gravity, a Mars-sized protoplanet rudely smashed into Earth. This sent a coil of debris into an orbit around our embryonic planet that ultimately coalesced to form the Moon. This interloping proto-world was dubbed Theia, an Ancient Greek goddess and mother of Selene, the Moon's overseeing deity.

Several simulations puzzlingly concluded that the Moon would be mostly made of Theia's wreckage,[5] not Earth's, as the Apollo samples indicate. But a recent study may have solved this paradox. Earth, right after its conception, was covered in a global ocean of magma. Everything was molten rock. And most of the so-called giant-impact models of the Moon's formation didn't take this into account. If they did, they would find that the resulting Moon is made of 70 percent Earth material, closer to what the lunar samples suggest should be the case.[6]

The debate rages on, and anyone claiming anything about the Moon's origin with any confidence is frowned upon. "Trying to piece together any process that occurred 4.5 billion years ago from a handful of rocks is definitely . . . tricky," says Dan Moriarty, who looks at strange lunar structures at NASA's Goddard Space Flight Center in Greenbelt, Maryland.

The giant-impact hypothesis isn't unreasonable. In the early days of the solar system, when planets were still being assembled, there were huge rocks crashing into each other all over the place.

But a simple explanation doesn't necessarily make it true.

In order to explain the structure of spacetime, Einstein had to lay out gravity's principles with reams of testable equations. It is all very elegant and pretty complicated—and definitely correct—but there is a lot of work still to do: scientists still haven't found the particle responsible for gravity's existence. "All Newton had to do was say gravity exists and he got all this fame because nobody had ever said it before," says Laura Kerber, a planetary scientist at NASA's Jet Propulsion Laboratory in Pasadena, California. "I kinda feel like that about planetary science. It's like, well nobody ever looked at this Moon before, therefore if you say very simple things about it then you're a genius!"

In other words, Theia is a decent, uncomplicated origin story for the Moon. But we can't say if it's correct or not until future work shows that to be the case.

Not knowing where the Moon came from is far from ideal. But thanks to the lunar sample's radioactive stopwatches, we know for certain that the Moon is 4.5 billion years old. Still, scientists aren't entirely sure what it looked like back then. Many reckon that it was also covered in a global ocean of magma shortly after it formed. Being 81 times less massive than Earth, the young Moon didn't have enough gravity to stop gases at the surface from escaping into space. With no atmosphere to shield it, the lunar magma ocean was exposed to the unfathomably cold tendrils of space and quickly began to freeze.

As it did, minerals in the ocean began to appear as crystals; some appeared earlier, others later; some rose to the top, others sank. Anywhere up to 100 million years after Theia's self-destructing run, most of the magma ocean would have solidified. Et voilá: you built yourself a Moon, still simmering from within, like a newly baked pie. "It's a story that's still being worked out," says Zeigler. But it's more or less true, he adds, based on the limited evidence we have.

A lot happened when the Moon cooled down. What matters is that, a bit like when you leave your soup to cool for too long, you get a gross crust at the top. You also get a lunar mantle beneath it; like Earth's, it is a squidgy layer that is pretty hot but mostly solid. There are radioactive elements in the mantle. As they decay, they let out heat, which melts other parts of the mantle. The somewhat melted mantle rises toward the crust and goes from a high-pressure environment to a low-pressure one. As on Earth, when you decompress bits of the mantle, it melts even more.

Ultimately, these fairly molten clumps keep on rising, and they push molten rock into the crust over time. Melt in the crust is better known as magma. And magma likes to do one thing above all else: erupt.

Strangely, the evidence for the oldest eruptions on the Moon wasn't found on the Moon. In September 1999, a gray-black rock was found sitting incongruously in front of a golden sand dune in the Kalahari Desert in Botswana. And in 2013, a dark green–tinged rock was found in Antarctica. These rocks, dubbed Kalahari 009 and MIL 13317, respectively, were identified as lunar meteorites. They were analyzed by Snape and his colleagues, and in 2018 they published a paper[7] listing the ingredients of these rocks. There were bits in MIL 13317 that were 4,332 million years old; those in Kalahari 009 were 4,369 million years old. Both are the products of the very first eruptions on the Moon. This means that straight away, as the lunar magma ocean transformed into solid rock, lava was rushing out into space.

JULES VERNE WROTE a couple of fantastical stories about the Moon, including his 1865 novel, *From the Earth to the Moon*, and an

1870 sequel, *Around the Moon.* They regale us with the efforts of a society of post–American Civil War weapon enthusiasts, the Baltimore Gun Club. This band of boom-stick proponents hopes to fire the club's president, his rival, and a French daredevil in a cannonball-like ship to the Moon. They ultimately succeed, albeit with many significant hiccups along the way.

At several points, Verne's characters speculate on the nature of the Moon's volcanoes. They describe "vast plains" that are "strewn with blocks of lava." They suspect the volcanoes are extinct, but had once fired meteoroids into space. Michel, the daredevil, rhapsodizes[8] about what it would have been like to see those volcanoes erupt, while pondering on their source of volcanic fuel:

> "Ah! my friends," exclaims Michel, "can you picture to yourselves what this now peaceful orb of night must have been when its craters, filled with thunderings, vomited at the same time smoke and tongues of flame. What a wonderful spectacle then, and now what decay! This moon is nothing more than a thin carcase of fireworks, whose squibs, rockets, serpents, and suns, after a superb brilliancy, have left but sadly broken cases. Who can say the cause, the reason, the motive force of these cataclysms?"

It's a vivid picture. But the reality of the Moon's volcanism is far more astonishing, and far more alien, than anything conjured up by Verne's exceptional imagination. And nothing is more extraordinary than the maria, those vast frozen seas of lava.

Many of the maria were named by the Italian priest–astronomer Giovanni Battista Riccioli in 1651, and beautifully so.[9] The Moon has so many seas—of Rains (Imbrium), Serenity (Serenitatis), Crises (Crisium), Nectar (Nectaris), and Clouds (Nubium), to name a few. Some have little bays, like Sinus Iridum, the Bay of Rainbows. Many have impressive dimensions; the Sea of Rains is 760 miles across, wider than France. But that's nothing compared to the Oceanus Procellarum, the Ocean of Storms, a truly colossal, curved flood of lava 1,800 miles across. These dark patches appear to sit within huge depressions in the land, carved out of that original, reflective, whiter lunar crust. As such, the original crust is referred to as the lunar highlands, the shores of which were once lapped by waves of red-hot hellfire.

It would have been a remarkable sight. Imagine standing anywhere along the Atlantic coast, but instead of water, all you can see, right up to the horizon, is molten rock, hissing at the dark, cloudless night sky above.

The maria were once thought to be made of sediments left behind by water or degrading mountains. But in 1949, astronomer Ralph Baldwin suspected they were trapped floods of lava. The Apollo missions proved him right,[10] finding that many of the sampled lava flows were anywhere from 3.1 billion to 3.9 billion years old. That means there was a roughly 800-million-year period of time where the Moon made these huge seas of lava. The question, of course, is how? I don't know if you've ever tried, but melting rock to make magma is tricky. You can't just chuck it in the microwave. Kīlauea never made more than lakes and rivers of the stuff. Making an entire *sea* of lava, then, is pretty damn impressive.

Melted clumps of commonplace minerals in the mantle, like pyroxene and olivine, are probably the source of the magmas that erupted so spectacularly onto the surface as lava. The ingredients are not that controversial. But the sheer volume of the lava seas and oceans suggests the cooking process itself was somewhat extreme, requiring the involvement of something titanic. And what's more dramatic than a series of huge meteorites slamming into the Moon?

"In planetary science, we love explaining things away with giant impacts," says Snape. It's not as prosaic as it may seem. Around 3.9 billion years ago, he notes, "the inner solar system was still being battered to hell." Massive, speedy meteorites were hitting the Moon with such might that they melted lunar rock upon impact. The age of that once-melted rock, something named impact melt (not related to the lava seas we call maria), tells scientists how old the impacts are. Apollo samples suggest the lunar basins in which the maria sit were carved out between 4 billion and 3.8 billion years ago, a 200-million-year-long game of planetary pinball known as the *late heavy bombardment*.

It's hard to convey the astonishing magnitude of these basin-excavating impacts with words alone. But eye-watering numbers can help. One of the youngest impacts on the Moon was made by a rock the size of New Jersey, weighing 25,000 trillion tons and moving at 52,000 miles per hour. The energy released in the blast would surmount, by many orders of magnitude, what could be unleashed by every nuclear weapon on Earth. Hitting at an angle, it sent debris bounding across the lunar surface, carving out gigantic grooves like bullets being fired into Jello. If you stood on Earth at the time, you would see a huge flash of light, followed by a cloud of dust and rocky projectiles blasting off into the great beyond. Meteors would cascade down through the planet's atmosphere, leaving flaming streaks in the sky. When all was said and done, the impact left a hole in the lunar crust seven times wider than the pit made by the asteroid that smashed into Earth 66 million years ago.

The fact that the seas are confined to these giant basins isn't a coincidence. Perhaps those impacts carved out conveniently located bowls that later lava flows could fill up.

But some wonder if the impacts cracked the crust, allowing magma, once trapped below but now liberated, to erupt into the basins. The impacts also removed so much of the crust at the point of destruction that it took a lot of pressure off the underlying mushy mantle. Those impacts excavated such a vast amount of the crust that, some suspect, there was *a lot* of decompression, much melting, and plenty of magma generation, leading to lava sea–forming eruptions.[11]

This sounds straightforward enough. But Michael Sori, a planetary scientist at Purdue University, says that this impact-and-decompression driven volcanism is a hugely controversial idea. "This is one of those things where I've seen scientists really go at each other in a vicious way," he says, evoking images of extremely curtly written email exchanges, dramatic fist shaking, and a lot of exaggerated eye-rolling.

The problem is that the maria turned up late to the party. They are often several hundreds of millions of years younger than the basins in which they appeared, meaning they erupted long after the craters formed. It is as if someone was shot, but the blood didn't pour out of the wound until a few months afterward. "It's like, what the *hell*, Moon?" asks Moriarty.

One issue is that the formation age of only a single basin, Imbrium, the one containing the Sea of Rains, is unambiguous. On the basis of samples taken from the shores of the Sea of Rains by the crew of *Apollo 15*, it clocks in at 3.9 billion years old. There were samples of other basins taken by other Apollo crews too, but they may have been picking up debris sent across the lunar surface by the massive impact that carved out Imbrium basin. It's possible that what we think are samples from separate basins may have all come from Imbrium. If that is true, then we don't know when the other basins formed. Without knowing that, we don't know if there really was a huge gap in time between the basins being carved out and the lava seas flooding into them.

Impacts and lava floods shaped the face of the Moon, and yet "we don't even know if they happened at the same time," says Moriarty. It's a bit like coming home to find your dog has destroyed the pillows on the couch, the toilet paper, the television remote, and a few books: you don't really know which of these fundamental acts of destruction happened first, or last.

Can the lava seas be explained by radioactivity instead? One of the last things to solidify from the magma ocean were rocks enriched in the goofily named KREEP material: that's potassium (K, on the periodic table), a bunch of odd elements dubbed "rare earth elements" (REE), and phosphorus (P). Several radioactive elements staying in the molten soup until the very end, like uranium and thorium, are clumped together with this KREEP material. Actively and excitedly decaying, these radioactive elements gave off plenty of heat. The energetic signal of thorium's radioactive decay could be detected from orbit. By mapping out the presence of thorium, scientists were able to find out where KREEP material appeared on the lunar surface.

The evolution of the Moon's magma ocean should have dumped KREEP material more or less evenly across the surface. But scientists found that it was concentrated in mostly one part of the Moon: inside and around the Sea of Rains.[12] In other words, heat-producing matter tends to appear close to where those huge lava seas are concentrated: on the nearside of the Moon. That link makes sense. More radioactivity equals more heat, which equals more melting, which equals more lava that spilled into the big impact basins nearby.

But why the KREEP material only expresses itself on the surface of the nearside of the Moon is anyone's guess. In fact, it isn't just the lunar seas and radioactive material that prefer the nearside. These differences are features of a larger lunar mystery: the nearside of the Moon looks *nothing* like the farside in almost every sense. And if we can't understand why, our quest to find out how the Moon made its maria will be lost.

OVER TIME, Earth's gravitational pull on the Moon altered our dance partner's rotation. The Moon eventually spun at such a particular speed that only one side, the nearside, always faces Earth. Presuming you aren't suffering from lycanthropy, I suggest you go outside and look at the Moon when it's full. See those lighter bits? They are the lunar highlands. The darker patches are the lava seas. You can easily see that the lava seas make up a significant proportion of the nearside.

We can't see it with our own eyes, but space telescopes show us an altogether different world on the farside: one full of craters, but largely bright and reflective, occasionally dusted by shards of crust and meteorites thrown up by impacts. The farside has maria too,[13] but the vast majority are found on the nearside. And the differences aren't just surficial. *GRAIL* data show us that the crust on the nearside is thin, sometimes half as thick as the crust on the farside.

"They basically look like two different planets," says Moriarty. And no one knows why.

I ask Tracy Gregg for an explanation during our video call, and she throws her hands up. Gregg, an expert on planetary volcanology at the University at Buffalo College of Arts and Sciences, would be a joy to watch lecture in person. She gesticulates with abandon and speaks of extraterrestrial volcanoes like she has only just found out about their existence. Every question I ask about the Moon's volcanoes is answered with animated, gleeful verbosity. Certainty is exciting. Uncertainty is perversely thrilling, too. When you talk to Gregg, it's a rollercoaster. She is the kind of university lecturer you wish you had.

With the exception of Mercury—"Mercury is weird," she assures me—the rocky worlds of the solar system have two major kinds of crust: Earth has oceanic crust, the stuff of seafloors, and continental crust, the stuff most of us live on. Venus has bright, mangled up mosaic-like ground, and everything else. But sometimes, one side of a planet is dominated by one crust type: Mars has lowlands in its northern hemisphere and mountains south of the equator. The Moon has its thin, lava sea–coated nearside and its pockmarked, thick farside. "Why is it so freaking asymmetric?" exclaims Gregg.

Like I said, if we don't have an answer to this question, we cannot hope to explain the Moon's lunar seas. So let's take a crack at it.

The crust being thicker on the farside could be explained by a really cool idea from 2014 named Earthshine.[14] Back when the Moon formed, it was much closer to Earth, perhaps only 8,000 miles away, compared with today's distance of nearly 240,000 miles. It would have loomed terrifyingly large in the sky.

Scientists wondered what effect Earth's magma ocean may have had on the newborn Moon—"before it was even the Moon, when Earth was still putting itself back together after the impact, and the Moon was still assembling itself," says Jason Wright, an astronomer and astrophysicist at Pennsylvania State University. Led by Wright's then–graduate student, Arpita Roy, a team of scientists noted the thicker farside crust has more of two key elements, calcium and aluminum. Such a major difference between the Moon's nearside and farside meant that "everything else was details," says Wright. It was, he adds, screamingly obvious: the farside "was in the shadow of Earthshine."

Presuming the Moon's nearside would have quickly gotten stuck permanently facing Earth, it would have been grilled by the extremely close and still molten planet. Much of the nearside would remain liquefied for a protracted amount of time.

As the Moon cooled, minerals containing aluminum and calcium would be among the first elements to solidify because of the chemical proclivities of these two elements. As the side of the Moon facing away from Earth's grill and into space was cold, these minerals preferentially crystallized out over there. That, according to Roy and company, is why the crust is so thick on the farside.

The idea that the young Earth turned the Moon into a crème brûlée "is a really cool concept," says Moriarty. "It's one of several possibilities you can't disprove." But even if that explains the thicker farside, it doesn't explain why lava seas preferentially sloshed about on the nearside. Like cream bursting through delicate filo pastry, the thinner crust of the nearside should allow for an easier route for those less delicious magmas to push through and erupt onto the surface. The thicker crust on the farside should impede that very same magma's upward exodus, which could explain the overabundance of lava seas on the nearside.

Except that it doesn't. The Moon's magma is rather dense, creating pockets of greater gravity. But, says Joy, the *GRAIL* mission didn't find any high-gravity spots beneath the farside crust that suggest magma got stuck in a subterranean traffic jam beneath a thicker crust.

As is de rigueur for planetary scientists, they've pointed the finger at yet another giant impact.[15] This time, the impact encompassed most of the nearside, leaving behind a gaping wound ripe for magma upwelling. This idea crops up every now and then. "It's always easy in planetary science to default to a giant impact when you encounter something you don't know," says Sori. It has been argued that the vast Ocean of Storms was once a big hole that lava erupted into, but *GRAIL*'s gravity data didn't demonstrate any circle or oval-shaped underground scars, the sort you would expect a huge impact to make.

There's nothing inherently wrong with invoking a giant impact, says Gregg. But you must make the physics work. And when you're talking about a hemisphere-size impact event, "there's a really fine line between an impact that's large enough to reface a hemisphere, and an impact that will completely obliterate the planet." Simple idea, but difficult in execution. We don't want to blow up the Moon.

And besides, if all it takes is a huge impact to make an ocean of lava, then what the hell is going on with the Moon's biggest maw? Around 4.3 billion years ago, a 105-mile-long space rock slammed head-on at 22,300 miles per hour into what is now the southern section of the Moon's farside.[16] Or perhaps it was a 124-mile-long rock moving at 33,500 miles per hour and hitting the Moon at a 45-degree angle.[17] But like debating the angle and speed at which a bowling ball hit a bystander's leg rather than the pins, these quibbles don't really matter as much as the result, and this impact was unquestionably ruinous: it created a 1,550-mile-long, 5-mile-deep impact crater. Named the South Pole–Aitken basin (it just barely overlaps the lunar south pole), you might expect it to be filled with its own lava seas, like the big craters on the nearside.

Nope. Although it certainly has its fair share of mare deposits,[18] it isn't flooded with them. One paper[19] suggests that so much of the crust was cleaved away by the impact that the underlying magma found itself too exposed to the chilly vacuum of space, freezing it up before it could have the chance to erupt.

There does appear to be a different sort of magma-like material in the enormous basin, says Moriarty, but it doesn't have the same chemistry that makes up the nearside's lava seas. Perhaps the impact may have blasted some of the geologic ingredients you need to make the "classic" maria into space, ruining the recipe.

It's another neat little story. But it doesn't change the fact that every single idea hoping to explain the differences between the Moon's face and its backside can't explain why the nearside has the lion's share of the lava seas.

You might be wondering whether *GRAIL* does anything other than act as a massive buzzkill for everyone's nice ideas. Why, yes! In 2014, it lent its support to one of the simpler theories. It found enormous scar tracks beneath the nearside.[20] Geologists think that these are huge rifts in the primordial lunar surface, chasms along which the crust was being torn in opposite directions as magma swelled below and huge volumes of lava came pouring out. If the impact basins formed atop those rifts, the lava may have simply flooded in from the sides and from below. Perhaps the impact basins are, at the end of the day, nothing more than bowls of lava-catching convenience.

But wait a minute. Why did all those rifts only form on the nearside?

You guessed it: no one knows.

With the samples we currently have, no one will ever know what made the nearside of the Moon so goddamn special. This asymmetry will remain a pain in everyone's collective ass, which means the origins of the lava seas will remain an unanswered question. "That story is still being written," Gregg says, grinning. "And the longer the story becomes, the clearer it is that we don't know exactly what's going on."

THE LAVA SEAS are arguably the Moon's oddest volcanic feature. But they are still just one stop on the grand tour of the lunar kingdom's volcanic wilderness.

Harrison Schmitt, the professional geologist flung onto the Moon during the *Apollo 17* mission, was bouncing across the lunar surface in December 1972 when he spied something unexpected. "Everything on the Moon is 256 shades of gray," says Gregg. "And he caught this flash of orange." He hopped over and scooped up some of the samples. They were orange beads of glass. And they weren't alone. Two years later, scientists perusing through the returned lunar samples found glass beads—black and green ones—in the *Apollo 11* and *Apollo 15* samples too, ranging from the size of human red blood cells to those the size of an ant.[21] Their origin was clear, and they told a remarkable story: billions of years ago, there were fountains of lava on the Moon.

Remember those lava fountains that emerged from Kīlauea's fissure 8 during its 2018 masterpiece? The same happened on the Moon, but in a distinctly alien fashion. Gregg, bristling with excitement, walks me through it.

Eons ago, gassy magma below rushed up through cracks, fissures, and vents on the surface. Water and carbon dioxide are the two most common gasses dissolved in magmas on Earth. But the Moon, being dry and lacking oxygen, probably went for carbon monoxide instead. "And there's no atmospheric drag and a sixth of the gravity," says Gregg. That means millions of small blebs of erupted lava cooling into volcanic glass could effectively glide for dozens of miles across the lunar landscape.

Glass shards of all shapes and sizes can be found all over the Moon, says Henderson. Some fields of volcanic glass cover nearly 400 square miles, about 17 times the area of Manhattan. The maria may not have an obvious source, but these glazy pastures sometimes do. A huge halo of volcanic glass can be seen inside Schrödinger crater. It surrounds a bump at the center, one that's 1,500 feet tall and covers an area double the size of Chicago. That bump has a hole punched into it. That hole is a vent, and that bump is a probably a big volcano that, long ago, blasted out fountains of lava, raining glass down upon the Moon.

You can find thousands of objects that look like volcanoes all over the Moon. Many are very likely to be volcanoes, but because this cannot be definitively determined from orbit, they are often given vaguer names, like bumps, domes, or cones. Marius Hills, found in the Ocean of Storms, is a collection of hundreds of domes and cones, each built by its own fountain of lava. Some have collapsed sides. On Earth, volcanic cones can have one of their flanks collapse due to structural weakness. This probably happened on the Moon too, but some of the larger lava blobs escaping the vent levitated in the low-gravity confines of the cone. From there, a blob could drift into the middle of the cone's wall and bore a hole through it. In other words, elongated blobs of lava once floated about above the Moon, causing havoc.

Although individually built, these pyric pyramids may all be connected underground. Lasers attached to the *Lunar Reconnaissance Orbiter* were used to carefully measure the distance from the spacecraft to the surface around Marius Hills. This work found that these cones and domes sit atop a huge volcanic mound, one whose incline was so small that human eyes couldn't perceive it. "If you were hiking on this with a full backpack, you still wouldn't notice you were going uphill," says Gregg.

This mound may be a volcano not unlike Kīlauea, and those many rotund rocky arenas may be parasitic volcanoes all feeding off the same big magma source that built the entire thing. "It's just weird," says Gregg. Lunar lava was runny like honey, and it would have taken something with the consistency of peanut butter to make a volcano this big.

And although cone collections can be found elsewhere, none have been found sitting atop a big rocky swell like a crowd of geologic zits. "You don't see this anywhere else on the Moon," she adds.

Headscratchers aside, everyone agrees that the aesthetics of eruptions on the Moon would have been mind-blowing. Whether it came from a dome, cone, fissure, or vent, if the Moon's fountains of eruptions were still happening today, you would be able to see any nearside glows of lunar lava all the way from Earth. "If you were in the middle of nowhere, you might be able to see the flash with the naked eye," says Gregg.

The Moon, lacking an atmosphere and, therefore, any way to trap heat on the surface, meant that the fountains' tiny lava blebs would have immediately frozen into a glass. With no wind to blow them about or any air to create friction, lunar glass beads are always perfect spheres.

Well, almost always.

Clive Neal, a lunar geologist at the University of Notre Dame, tells me that a student of his was picking through some of the orange *Apollo 17* glass beads when he found one that was shaped like a tear. On Earth, this would be referred to as a Pele's tear. "You can only get a Pele's tear if it is shaped by falling through an atmosphere," Neal says, which means that, once upon a time, the Moon had an atmosphere—a truly alien thought.

In 2017, scientists took a closer look at the maria and tallied up the ancient gases trapped within the frozen lava blobs. They reckon that the prolific outpouring of lava that made the maria would have brought trillions of tons of carbon monoxide, sulfur, and water vapor along for the ride. If enough of those gases escaped skyward fast enough, they could have built an atmosphere around 3.5 billion years ago. It would have been very thin compared to Earth's, but it may have persisted for 70 million years.[22] The Moon would have had gusts of wind, perhaps even ephemeral clouds. According to Debra Needham, a planetary scientist at NASA's Marshall Space Flight Center in Huntsville, Alabama, and one of the authors of this research, there may have once been precipitation on the Moon: rain, over the Sea of Rains.

Not all of those gases would have escaped into space. "The permanently shadowed regions at the poles are some of the coldest places in the universe," says Moriarty. If water molecules flitted over to these shadowed pockets, they would have become trapped there forever as they frosted out as ice.

There were rivers, too. From orbit, it's easy to see that much of the Moon is garnished with meandering channels, often starting from a volcanic vent but seemingly disappearing into nothingness at the other end. There is no splaying out at the end, no river delta; the channels don't bifurcate into distributaries, nor do they flow into lakebeds.

"Oh, these are my favorites!" Gregg says, clutching her hands, clearly delighted. These channels are named sinuous rilles, and they were made not by water, but lava. "But lava on Earth doesn't do that," she says. It doesn't wind about quite so erratically, nor do lava channels just up and vanish. These squiggles are also impressively lengthy, extending for hundreds of miles from their source.

There are two diametrically opposed models that hope to explain the Moon's rilles. First, the tortoise: these were once lava tubes, just like the ones in Hawai'i. Lava is flowing underground beneath a frozen lava roof. The molten rock is kept insulated and hot, so it flows for a considerably long distance. On the Moon, these flows were very slow, filling up one Olympic-size swimming pool every four minutes. They lasted decades before losing their magmatic mojo. The lava drained out, and the unsupported roof collapsed, leaving the rilles.

Then, the hare: the lunar lavas are actually hotter than the hottest lava on Earth by a couple of hundred degrees, flowing like rapids in a river, enough to fill up anywhere from 40 to 400 Olympic-size swimming pools every single second. Being so hot and turbulent, they forcefully eroded out the channels beneath them.

"Like so many things in geology, the answer is likely to lie somewhere in-between," says Gregg—although if the hare wins out, it suggests something truly strange was happening in the Moon's magma factories.

Scientists came tantalizingly close to finding out the answer during the *Apollo 15* mission. The crew were asked to jump down into one of these sinuous rilles in order to grab a tell-all sample. "These guys were kind of the epitome of machismo at that time," says Gregg. "But they got to the edge of this rille, looked down, and said: 'no f**king way, that's way too steep.' So you kind of have to believe them!" I suppose it would have been a bit awkward if an Apollo astronaut died after getting stuck in an ancient lava channel. "Just a bit awkward, yeah," she says, smirking.

Whichever is the right answer, lava definitely snuck around below the lunar soil. Gregg, sharing her screen, wobbles her cursor around a series of pits snaking about near a sizable crater in the Marius Hills region. "See those?" she says. "Those are holes in the tops of a lava tube." Cavernous tunnels of volcanic rock where lava once flowed. "A lunar cave where you could shove some astronauts," says Gregg—not for nefarious purposes, but in order to shield them from dangerous cosmic radiation bombarding the lunar surface. Some suspect that, in the future, these labyrinthine networks of tubes could be great places to build shelters for astronauts staying on the Moon for the long haul. NASA's Kerber, keen to sample the layers of solid lava concealed within, wants to send a spelunking robot into one of the Moon's lava-tube skylights.[23] But, facing steep competition from other teams of scientists with their own mission proposals to explore the solar system, she hasn't convinced the higher-ups that it's worthy of funding just yet.

IF, ON A LUNAR SAFARI, you aren't content with seeing lava caves, vistas of volcanic glass, and hills of ancient fire, I would wholeheartedly advise you to pay a visit to Ina D: a letter D–shaped depression on the summit of a mound the size of a large city, itself sitting within Lacus Felicitatis, the Lake of Happiness. Inside the volcanic crater–like depression you can find bulbous mounds of what looks like pools of mercury.[24] They were first spotted from orbit by *Apollo 15* in 1971.

"Ohhh," says Gregg. "Those are weird." These dark, cratered mounds look a bit like tiny lunar seas, so they are definitely volcanic. But their amorphous nature and the lack of consistency between mounds defy explanation. Some suspect that they are lava flows that got stuck and then inflated by the injection of more lava from below. Others wonder if they represent lava oozing out of cracks in a capped, cooling lava lake: in a low-gravity, atmosphere-lacking environment, the bubbles of gas trapped in the lava would have been able to dramatically expand, creating big foamy lava mounds[25] that then froze, like an avant-garde take on a cappuccino. But their shapes and possible origin stories aren't the strangest thing about them. Their ages are.

These bizarre mercury-esque blobs are part of a family of strange patches of frozen lava found all over the lunar nearside. They can be smooth, rough, elongated, round, darker, brighter, plentiful, or isolated. They are, rather helpfully, called irregular mare patches, or IMPs. And in 2014, a team of scientists carefully used crater counting to determine their ages. They found that they were no more than 100 million years old,[26] meaning that dinosaurs should have been able to see eruptions on the Moon. That may sound ancient to you and me, but for the Moon, 100-million-year-old volcanism is practically yesterday. And that could be a problem.

The lava seas suggest the bulk of lunar volcanism happened between 3.9 billion and 3.1 billion years ago. But eruptions still took place long after that epoch. In the South Pole–Aitken basin, lava patches of all ages can be found, from 2.2 billion years old to perhaps as young as 1.6 billion years old. Some lava deposits in the Ocean of Storms appear to be a mere 1.2 billion years old.[27]

Crater counting isn't infallible. On its own, it only tells us whether one surface is older or younger relative to another surface. The ages of the limited number of lunar samples we have can be precisely determined when their radioactive stopwatches are checked in laboratories. These ages are then matched up with crater-counted lunar surfaces in order to estimate those surfaces' actual ages. But such ages are still estimates, not definitive values.

But presuming that IMP-hunting study is right ("I think it's a well-constructed argument. It's hard to refute it," says Gregg), and 100-million-year-old eruptions did happen on the Moon, then we really have no idea what is going on.

The Moon, compared to Earth, is small, about 2 percent of the planet's volume. You need heat to melt things in order to get eruptions, and for Earth and the Moon, that comes in two forms: the leftover heat trapped in their hearts from their violent formation, and the decay of radioactive elements. Larger worlds keep their heat trapped in for longer.

Our miniature Moon, then, should have gone cold long ago. Its mushy mantle should have seized up, and its magma supplies should have run dry. Its most prolific volcanic period was more than 3 billion years ago, when its innards were still scorching. You would expect localized eruptions for some time afterward as pockets here and there remained hot. But how could the Moon's viscera have remained so hot for so long that a *Stegosaurus* could have seen lava fountains paint the night sky?

A fundamental problem is that we don't know whether the Moon is hot or not. And if its hotness is debatable, we can't say how it could trigger youthful eruptions. "It's controversial to say that something was happening 100 million years ago," says Joy. "It's still controversial to say that something was happening a billion years ago."

Some models use the amount of lava the Moon erupted in the past to work out how hot the Moon has been over the past few billion years. That's fair enough: more lava roughly equals more heat on the magma-making barbeque. But it turns out that we haven't found all of the Moon's maria. Impacts sending debris across the surface, as well as younger lava flows, have buried old lava seas beyond sight.

Sori is one of the scientists looking for these so-called cryptomaria. Recent impacts sometimes blow the lid off the sarcophaguses of concealed maria, leaving behind telltale darker patches around their lunar wounds. But being old, cold, and dense, hidden lava seas can also be spotted using *GRAIL*'s gravity data. Not long ago, Sori and his colleagues found suspicious gravity spikes that could be buried maria all over the Moon.[28] The largest area of potentially submerged frozen lava was found forming a crescent shape, hundreds of miles long, arcing over the southern lunar nearside.

But even with the Moon's invisible volcanic grin, they worked out that there may only be a modest amount of hidden maria on the Moon. There was nothing to suggest it has been anomalously hot, and certainly nothing informative enough to indicate how long the Moon has remained capable of making magma over billions of years.

Another dead end.

Scientists are aching for samples of the Moon's mantle, that hot malleable slice of the Moon that sits under the crust. If a sample is found, says Joy, we could use it to find out what the Moon's magmatic kitchen is like, to raid its pantry and find out what ingredients and cooking utensils it has been using to bake its batches of magma. It would go a long way toward telling us how hot things would need to be to produce volcanism over billions of years. But, to date, the Moon's mantle has evaded capture.

Despite being there not too long ago, we also don't know how hot the Moon is now. Several thermometers[29] were buried in the lunar soil by the *Apollo 15* and *Apollo 17* missions and kept running from 1971 to 1977. It was hoped they would take the Moon's internal temperature, but as they only went a few feet into the ground, they were affected by whatever was happening at the surface. Scientists found that they registered a curious uptick in temperature during the duration of the heat-flow experiment. This was found to be the astronauts' fault: as they bounded about, they kicked up a lot of lunar soil. This made the surface rougher, making it less able to reflect sunlight, and causing the surface to warm up.[30]

Seismometers left behind by the Apollo missions—switched off in 1977, like the thermometers, because of budget cuts—discovered moonquakes. They never get as strong as earthquakes, but they last a lot longer. On Earth, a magnitude 5.0 quake, enough to shake buildings and cause damage, usually lasts for a matter of seconds. On the Moon, says Neal, they can last for 10 minutes, making them an important consideration for future human outposts.

Like earthquakes, moonquakes happen on faults. But unlike terrestrial temblors, moonquakes are largely driven by the Moon cooling down. As it continues to lose its primordial embers, the entire Moon shrinks, a bit like a grape turning into a raisin. This global squashing causes faults to slip, especially those given a little extra tug by Earth's gravity.

If the Moon's cooling is largely complete, it should stop contracting and quaking. But Apollo-era moonquakes suggest this isn't the case. And a recent study[31] found that many of the faults responsible for the Moon's recent quakes aren't reactivated old ones, but youthful cracks in the ground.

The Moon isn't warm by anyone's standards. It is cold enough now that it is very nearly a planetary corpse. But the Apollo-era moonquakes suggest it remains warm. And the IMPs mean that it was still somehow hot enough to make volcanoes erupt into space yesterday, geologically speaking. "I would not be the least bit surprised to find out that there are tiny little spots on the Moon that are still cooking," says Gregg.

Sensors on spacecraft have detected slight changes in the whispers of gas flitting about above the Moon. Perhaps there are still pockets of molten magma here and there, blowing off some steam. The Moon may have entered retirement, but in a gassy, indigestion-like respect, it is alive and kicking, having a jolly old time, says Joy.

But did these pockets recently erupt? No one knows, partly because scientists don't even know if magma pockets actually existed on the Moon. "We don't even know whether magma moving up through the crust gets ponded in magma chambers or whether they just whizz up," says Joy. This is not a great position to be in if we hope to understand the Moon's volcanoes. It is like asking a car mechanic how a car works and the mechanic sheepishly telling you that they aren't even sure if cars have fuel tanks.

Where did the Moon come from? Why does it have two completely different faces? What made its seas of lava? Do volcanoes on the Moon behave like Earth's? How long was it erupting? With a mixture of excitement and frustration, Joy summarizes humanity's cumulative understanding of our planet's celestial companion: "We don't know the answers to any of these questions!" she exclaims. From start to end, the Moon's history is an unsolved, broken, partially invisible jigsaw puzzle. We don't know where all its pieces are, and we don't know what it's supposed to look like when it's assembled.

IMAGINE YOU ARE ON a tour of a famed historic or natural locale, and every time you ask your guide a question, the guide shrugs in response. By the end, you'd probably think that was a pretty crappy tour. But it isn't always the tour guide's fault. Sometimes, scientists don't know as much as you demand—and they are fully aware of it. Science isn't about certainty. Research doesn't always come down to a Poirot-style dramatic reveal at the end of the story. It's about pushing back the shadows of uncertainty, bit by bit.

Gregg tells me that many of her students think that the Moon is a bit like a completed video game, with all its story and secrets revealed to the players. Some of them actually think humans have already been to Mars, because how could we not have been half a century after landing on the Moon? But a handful of rocks and a small fleet of orbiting spacecraft aren't enough to crack the case of the pale guardian. We still know so little about it, especially when it comes to its volcanism. And without going back to continue our investigation, she says, Earth itself will always remain a somewhat alien planet.

Earth, as you may have noticed, is covered in erupting volcanoes. It also has so much more: continents, oceans, seafloors, mountains. It is a kaleidoscopic dream of life-supporting geology. But in the beginning, it was covered in a magma ocean, just like the Moon. Both worlds evolved very differently as their molten rock oceans froze over. The Moon shows us a fate that could have easily befallen Earth if things deep below, processes we are still unpacking, operated a little differently.

Earth is excellent at burying its past: its wind, rain, seas, and oceans eat away at exposed rock, while large parts of Earth's crust get destroyed over time as it sinks down into the annihilative depths of the planet. Our planet's formative days, which could be recounted by the first generation of rocks it made, have been erased from existence. But the Moon could help Earth remember its past. Scientists largely agree that the Moon was formed when something cataclysmic happened to Earth in its youth—most likely Theia slamming into it. As a result, plenty of the Moon's geology was sourced from Earth itself. That means rocks as old as the Earth itself, made from the planet's own matter, hide in the Moon's volcanic vaults, untouched and uncorrupted on a world lacking air or water. Finding these rocks will allow us to see Earth as it was when it came screaming into the cosmos.

Volcanic rock is fantastic at trapping secrets, and the lunar samples we already have hint at the revelations to come during future missions to the Moon. Nearly everyone thought Theia's impact and the subsequent global magma ocean would have easily boiled away most of the Moon's water and its easily vaporized elements, such as fluorine and chlorine. This idea was so set in stone that it took researchers several years to actually convince people to fund their study to double-check this by looking through various Apollo samples. They succeeded, and in 2008,[32] a new scientific technique that enabled us to effectively see inside those volcanic glass beads found that they were imprisoning ancient water. And in 2017,[33] India's lunar orbiter, *Chandrayaan-1*, identified plenty of hydrated volcanic glass all over the lunar surface, suggesting Moon's deep interior was seriously soggy.

That doesn't mean the giant impact and global magma ocean didn't vaporize much of the Moon's original water. But a 2016 paper[34] suggested that a squadron of water-rich asteroids arrived on the scene a little later on and dive-bombed the Moon, rehydrating its innards. Theia's impact on a young Earth would have stripped away much of its water too, so perhaps we owe our seas and oceans, and by extension our very existence, to the violent visitation of damp asteroids from afar.

"Everything we learn about the Moon isn't just about the Moon," Joy says. What we find up there teaches us about, well, everything—from the beginning of time to the end of worlds. Unravelling its volcanic history won't just reveal some of the solar system's long-kept secrets. It'll teach us to ask better questions about the universe.

Our education has, sadly, stalled ever since America "won" the Space Race and political interest in the Moon's exploration vanished. Times are changing, though. China is on the march. Its *Chang'e-4* mission has been looking for mantle material in the South Pole–Aitken basin since January 2019. It's a cool little lander–rover pairing, and the first explorer to traverse the Moon's farside. Although these droids can only do a modicum of science with their onboard gizmos, their selenocentric shenanigans continue to impress. In November 2020, China sent its more daring successor, *Chang'e-5*, to the Moon—this time, to the Ocean of Storms—one that scooped up some rocks and returned them back home before the year's end. These samples, which will be comprehensively picked apart in high-tech laboratories, are humanity's first pristine lunar samples in more than four decades.

Not keen on losing political ground to its rival, and acutely aware that there are untapped mineral resources, including water-ice, in shadowed pockets of the Moon, America is now hoping to put astronauts back up there and begin the process of maintaining a long-term lunar presence. It may not be driven by the urge to learn more about the Moon, but science is going to happen regardless.

It's a new paradigm, says Kerber. You used to have to fight tooth-and-nail to justify conducting science on the Moon. Now, it's more of an invitation: we're going to the Moon, so do you want to come and do some science while we plant our flags?

The sequel to the Apollo program is named Artemis, named after Apollo's twin sister. The plan, as of 2020, is that following on from two test flights, *Artemis III* will land the first woman on the Moon in 2024, ideally somewhere near the lunar south pole.

Even if that 2024 deadline is missed, says Henderson—and many suspect that will indeed be the case—there will one day be lunar astronauts again. "There should be. There has to be," she says.

Dexterous, intelligent humans are not only capable of conducting far more science than their mechanical cousins. They are also able to bring far more samples back to Earth per round-trip mission. Astronauts would also be able to set up a geophysical network on the Moon that uncrewed spacecraft could manage. This would allow scientists to take the Moon's temperature, see down into its depths, look for relics of that magma ocean, and find out how active the Moon remains today. Zeigler looks forward to expanding NASA's lunar vault in the coming years. "These samples will unlock a whole new generation of lunar science—of solar system science," he says.

The Moon's volcanic rocks preserve all kinds of untold stories, and there is only one way we can read them: strap ourselves to rockets, point them at that beautiful milky orb, and go.

VI

THE TOPPLED GOD

Paul Byrne would love to blow up Mars. A planetary scientist at North Carolina State University, Byrne is one of my favorite scientists. He's blunt and funny, which makes him a joy to interview. We agree that things on Earth could be considerably better right now, and that other planets are awesome. But I do disagree with him about Mars. To be fair, he has wanted to blow up the Moon, too, and Jupiter and some of its moons. "I like to think I'm an equal opportunity blower-upper," he says. Part of this is just to wind other scientists up. Sometimes, a lot of attention is paid to a particular place in the solar system. It gets hyped up, for better or worse. The suggestion that said place should be blown up upsets a certain subset of overly serious scientists, and that gives Byrne a modicum of mischievous energy.

Mars undoubtedly gets targeted by his planet-killing ire most frequently. This may seem strange, considering Byrne did his PhD on martian volcanoes. Several of his grad students are working on Mars projects. If he was in charge of NASA, he wouldn't stop them from studying Mars because it is a treasure trove of scientific secrets. "But," he says, "Mars has got an absolutely disproportionate amount of attention over the last few decades."

It is true: many robots have been thrown at Mars—mostly by

NASA and the European Space Agency, with many more to come. There are a range of reasons for this: popular culture loves Mars, making it so engrained in the public psyche that space agencies can more easily justify chucking robots at it; an increasing tally of successful missions to Mars requires more advanced follow-up missions to answer increasing tantalizing questions, particularly with regard to the possibility of alien microbes; these funded missions fund scientists and their students, creating a positive feedback cycle of Mars-centric research; and Mars, compared to distant Mercury or the metal-eating environment of Venus, is easier to reach. The rest of the solar system deserves more attention, Byrne reckons. If he blew up Mars, then other planets and moons would get more time in the limelight.

But he should abandon his wicked plot to destroy Mars because of something NASA's *Mariner 9* spacecraft saw as it began orbiting the planet on November 14, 1971.[1] As it peered down into the orange-red dust, it spied four dark spots just south of its equator in a region known as Tharsis, the word for a Biblical land at the western edges of the known world. On closer inspection, nests of craters could be seen at the top of these dark spots, suggesting they were volcanoes—mountains of immense size, covering areas larger than entire countries with peaks beyond Everest's reach. They were so tall that even when the entire planet was smothered by a dust storm, *Mariner 9* could see their caps protruding through the rusty tempest.

Over time, and with successive orbiter missions, scientists began to realize that there wasn't just this quadruplet of volcanoes—the trio, collectively known as Tharsis Montes, and Olympus Mons sitting just off to the northwest of Tharsis—that showcased Mars's volcano-building prowess. The entire region of Tharsis, covering an area of nearly 12 million square miles, was itself a volcanic construction. Tharsis emerged so quickly, and became so massive, that it deformed the entire planet, warping the crust as if a cannonball was dropped on a swimming pool full of custard.

Despite being six times smaller than Earth by volume, Mars was somehow able to build the biggest volcanoes known to science, dwarfing any other planet's attempts, while keeping the eruptions going for 4 billion years—almost as long as the solar system is old. It is, without a doubt, a planet punching well above its weight.

Alas, Mars is not what it once was. There used to be an abundance of ice and water flowing all over its surface. Today, it is a frigid desert. Its atmosphere, once prominent, was destroyed. Radiation bombards the surface. Life may teem beneath the surface, but it would stand little chance of surviving should it venture up to see the stars.

Whether you're an eschatologist or an aficionado of comic book supervillains, the end of the world features heavily in our storytelling. Take your pick: nuclear war, an out-of-control artificial intelligence, an asteroid, a deadly pandemic—we've dreamed up all kinds of oblivion that could afflict us. But we're lucky. Earth, so far, remains a place where life can flourish. For Mars, the end of the world has already happened—and a clue, held within its volcanic vaults, may tell us why.

MARS IS A BIT LIKE Earth's cousin: it has some similarities and a whole lot of differences.

A day on Mars is 24 hours and 37 minutes. Both spin around on an axis, and each is tilted to a similarly moderate degree on its axis, so as Mars orbits the Sun once every 687 Earth-days, it also experiences seasons. Its surface is also made of rocks, many of which have been rusted red thanks to the plentiful iron that reacted with the atmosphere's oxygen. As with Earth, scientists suspect that below that crust is a squishy mantle and, deeper still, an iron-rich core.

It has an atmosphere, but it's 96 percent carbon dioxide, there isn't much of it, and it's not great. As you increase your altitude on Earth, the air gets thinner. This is partly because gravity keeps much of the air close to the ground, and most of the ground is at sea level; it is also partly because there are fewer gas molecules packed into the air at altitude. At the top of Everest, the air is so thin—or, to put it another way, the air pressure is so low—that you begin to suffocate if you stay up there for too long. It could be worse: there's so little atmospheric stuff on Mars that the surface air pressure is 50 times lower than at the roof of Everest. A decent amount of Mars's carbon dioxide is locked up in its polar ice caps, which shrink and expand during its respective summers and winters. Carbon dioxide dislikes being liquid, so although it doesn't rain on Mars, it does occasionally snow dry ice.

There's also less gravity on Mars—two-thirds less, thanks to Mars being just one-tenth the mass of Earth. Venus and Earth are roughly the same size, but Mars is anomalously small. It's not clear why, although one idea frames Jupiter as a thief. Shortly after the beginning of the solar system 4.6 billion years ago, large rocks dozens of miles across began clumping together under their own gravity. When enough of them are squashed up, you get what's called a planetary embryo. Normally, embryos combine with others. "You put 10 together to get Earth, nine together to get Venus," says Byrne. Remaining embryos are often sucked into the Sun or ejected from the solar system. Mars appears to have remained in its embryonic state, says Byrne, probably because "there's not just that much stuff left in that area of space, because by that point, Jupiter had formed." Thanks to its enormous proportions, Jupiter's gravitational pull is immense. "And whoosh!" he says, imitating a cosmic vacuum cleaner stealing the other planetary embryos that should have gone to Mars.

Mars has not one but two moons: Phobos and Deimos, ancient Greek for "fear" and "dread," respectively, apt names for the companions of Mars, the Roman god of war. But these Lilliputian space potatoes don't inspire either emotion. They are tiny, just 17 and 9 miles across, respectively. They also look like asteroids captured by Mars's gravity, but they orbit in a way that suggests Mars made them out of orbiting material shorn off from its surface after a giant impact. They also have opposite fates. Phobos is falling into the planet where, in the next 100 million years, it will be torn apart by martian gravity and turned into a set of rings not unlike those adorning Jupiter, Saturn, Uranus, or Neptune. Deimos, on the other hand, is making its gradual escape out into the great shadowy beyond.

Mars, like the Moon, also has two hemispheres that are vastly different from each other: a southern, highly cratered highlands section with crust up to 62 miles thick, and a northern, smooth lowlands section with a paltry 19-mile-thick crust. Like the Moon, it really does look like the hemispheres of two different planets have been accidentally glued together by a deific architect on the third night shift of the week. No one knows for certain what happened here, but they normally involve—yep, you guessed it—giant impacts. One or many meteorites preferentially slammed into the north, carving out a hemisphere-sized crater.[2] Alternatively, a giant meteorite crashed into the south, creating a magma ocean that slowly cooled and baked up a rather thick crust.[3] Who knows, really. The origin story of this hemispheric dichotomy could come down to something far more peculiar than anything anyone has imagined yet.

And, of course, Mars has volcanoes—a proverbial zoo of them, in fact. Tracy Gregg, the effervescent planetary volcanologist from the University at Buffalo College of Arts and Sciences who took us on a volcanic safari of the Moon, understandably loves Mars too. You name it: if it involves magma, Mars probably has it.

Like the Moon or Earth, much of Mars's volcanism uses flavors of basalt, that relatively runny magma and lava that prefers to make lava flows rather than any sort of mega-explosions. You can find volcanoes all over Mars, from the cratered highlands around the giant Hellas basin impact scar in the south to Elysium Planitia, which sits just north of the martian equator. Lava flows are all over the place and can easily cover hundreds of thousands of square miles. Paterae—crater-like objects with scalloped edges—look a bit like the calderas atop Hawai'i's volcanoes. And, like Hawai'i, there are many shield volcanoes on Mars too: volcanic mountains built up by countless successive eruptions of lava that are many, many times wider at their base than they are tall, with gradual, barely there slopes leading to the summit. You can also spy all kinds of domes and cones too, sometimes appearing in the hundreds on volcanic plains or hiding out in the summits of the volcanoes themselves. They are sometimes parasites that stole magma from the huge reservoirs responsible for those larger shield volcanoes. Sinuous rilles, those anastomosing lava channels we see on the lunar surface, can be found slithering across Mars too, often from vents at the tops of volcanoes. There is ample evidence of magma coming into contact with water and ice too.[4,5,6,7] Several ash-encircled craters and cones look a lot like soggy explosion craters, the consequence of magma rapidly boiling water or ice under pressure.

There are even some places on Mars that look like the remains of pyroclastic flows, including chaotic, loosely packed rubble-like piles varyingly close to various volcanoes. The most promising are known as the Medusae Fossae Formation, a widely dispersed collection of debris that was dumped atop the land and was easily eroded by Mars's weak winds. The problem is that if they were emplaced by pyroclastic flows, we can't find the source volcano responsible. Some have suggested Apollinaris Mons to the west, but, says Gregg, "the hole is not big enough" to have produced such voluminous dragon's breath deposits.

Mars, like Earth, also has mud volcanoes. They are exactly what they sound like: buoyant or pressurized mud bubbles up out of holes in the ground. Terrestrial mud is fairly runny. But Mars's thin atmosphere means that the average surface temperature is –81 degrees Fahrenheit. This quickly freezes the tops of these mud flows, insulating the mud below and letting it flow over long distances like gloopy lava in tunnels in Hawai'i.[8]

David Bowie would've been happy to hear that there are spiders on Mars too.[9] When the Sun shines on the polar ice caps during martian summers, carbon dioxide ice trapped underground transforms into a gas. Needing a lot more space to be accommodated, this gas zips up and blasts martian soil skyward, producing black-ish specks that look like little spiders when viewed from orbit. The spider-covered land is technically known as araneiform terrain, a fine piece of scientific nomenclature that proves planetary scientists have the best names for things.

Mars, then, is an embarrassment of volcanism. It doesn't appear to be erupting today, but its storied past featured so many eruptions that the crust is one huge amalgam of lava. "Volcanoes are really the canvas for everything else that happened on Mars," says Frances Rivera-Hernández, a planetary geologist at Georgia Tech. "You originally started with volcanoes making igneous rocks, and everything else that's happened since then has been painted on that."

But all of this volcanism pales into near insignificance when you peer at Tharsis. It's a messy magmatic cake, built by countless eruptions onto and into the martian crust. Thousands of volcanoes, sprinkled atop a rocky rise, are somewhat buried by a generous pouring of lava icing. Like decorative figurines, several huge shield volcanoes, including the Tharsis Montes trio, adorn the uppermost icing layer. Undoubtedly Mars's volcanic masterpiece, it would triumph in the Great Cosmic Bake-Off.

"The magnitude of Tharsis, and even the individual volcanoes on or near it, like Olympus Mons . . . there's just nothing comparable on Earth," says Michael Manga, a geoscientist at the University of California, Berkeley. "If you look at a globe of Mars, you can easily see the volcanoes. If you do the same thing with Earth, you wouldn't see any volcanoes because they're too small."

Forget, for now, the volcanic titans that sit on the Tharsis rise or on its peripheries. The Tharsis rise itself is an extravagance of dimensions, looming above the martian surface like an unholy red balloon. Earth's largest volcano is, scientists suspect, Mauna Loa, a Hawaiian shield volcano that covers 2,000 square miles at its submerged base. Sure, that's big, coming in at roughly four times the area of the City of Los Angeles. But the Tharsis rise is more than three times the size of continental America.

It is an aberration, and scientists want to know how diminutive Mars was capable of creating it. The problem is, much like the Moon, scientists don't know much about Mars at all.

MARS IS POPULATED BY ROBOTS—those that land and stay put, those that rove across the surface, and those that linger in orbit above. About half of the missions to Mars have failed,[10] but those that have managed it have built on the revelations unearthed by their predecessors. Almost every part of the martian planetary system has been poked at, from its atmosphere to its magnetism, its geology to its water content, its topography to its geography. I'm writing this chapter during the so-called Summer of Mars, where three missions are launching to the burnt-orange orb: China's *Tianwen-1* (meaning "questions to heaven"), a planetary orbiter, lander, and rover combination; the United Arab Emirates' *Al-Amal* ("hope") orbiter; and NASA's *Perseverance* rover.

Perseverance is arguably the most scientifically important of the three. Instead of being a stand-alone mission, this rover will, among other things, collect dozens of samples of the martian surface and place them in sealed tubes all over Jezero Crater, its landing spot. A subsequent series of spacecraft, varyingly managed by NASA and the European Space Agency, will then aim to retrieve these samples.[11] A smaller rover, partly driven by an artificial intelligence, will collect these tubes before loading them in a rocket that will autonomously blast off the martian surface. The samples container will then be released into orbit around Mars, where an orbiter will catch it, like a glove catching a baseball, before sealing it up tight and flying it back to Earth. If all goes according to plan, the samples will make a hard landing in an American desert, where they will be snapped up and taken to the most secure laboratory on Earth. The idea is that these samples, the first pristine chunks of Mars—in other words, the martian equivalent of the Apollo samples—may contain life, so they need to be examined in a facility that guarantees any extraterrestrial life-forms won't find their way to the outside world.

The Mars Sample Return mission is fantastically ambitious. Any stumble along the way, be it technological or economic (more than two dozen countries have to keep funding it over more than a decade), will result in failure. But getting our hands on those rocks will tell us so much more about Mars than can ever be ascertained from orbit.

That said, we have already gotten our mitts on pieces of Mars. On October 3, 1962, a farmer near Katsina, Nigeria, was minding his own business in one of his cornfields when a large thud sent dirt and dust flying. A 40-pound rock had fallen from the sky and dug out a crater just 10 feet or so from where he was standing. Zagami, as the interplanetary interloper was dubbed, was—and still is—the largest martian meteorite ever found, and one of the few to be witnessed tumbling to Earth.

It wasn't clear at the time that this was a martian meteorite. But in 1976, two NASA Viking spacecraft touched down on Mars and took samples of the atmosphere. In 1983, scientists carefully looked at the gases trapped within the glassy pockets of a suspect meteorite named Elephant Moraine 79001 and found that the mixture was unmistakably martian.[12] Today, there are around 165 known martian meteorites. By mass, it is similar to the lunar meteorite collection we have. "So, it's not a huge amount," says John Pernet-Fisher, a planetary scientist studying the chemical evolution of worlds at the University of Manchester.

Martian meteorites are unquestionably valuable: their crystals contain the chemistries of lavas; their surfaces tell us whether or not water chipped away at them; their glasses seal up bubbles of ancient martian atmosphere that we can compare to today's thinner haze. But the three major suites of rock that martian meteorites encapsulate can't tell you about an entire planet. And they aren't pristine, having been bombarded by cosmic radiation and scorched by Earth's atmosphere.

Meteorites also have no context. "One of the biggest things we don't really know about is where they are from," says Pernet-Fisher. "Unlike the Moon, there's no way really of pinning it down to a particular geography. We just have this collection of volcanic rocks, and a couple of sedimentary rocks, and where exactly they're sourced from is a bit of a mystery." The chemistry of the meteorites certainly gives us clues, but despite several attempts[13] to unravel their provenances, "it's near on impossible to try and pinpoint specific craters to specific samples."

That huge question mark means that we can use the radioactivity of these meteorites to work out how old they are with remarkable precision, but we can't do the same for the martian surface. Just like on the Moon, the number of craters on a surface is used to work out how old a surface is relative to another, with more craters equaling an older surface. But on Mars, both in the past and today, its atmosphere, volcanism, ice, and water have eroded or buried many of these craters. Even when a surface has a seemingly reliable crater collection, the absolute ages of these surfaces—as in, the actual numerical ages in years—is estimated on the basis of what we know about the Apollo-sampled lunar surface. And we don't know much about that either.

Tanya Harrison, a planetary scientist and the founder of Professional Martian LLC, a company focused on science and sci-fi consulting, sums it up best. When it comes to the absolute ages of things on

Mars, "it's all guessing." Like the Moon, scientists' understanding of what's going on comes down to models. And so many models sound convincing because, says Manga, we have a very limited perspective: meteorites and robots. Models, whether simulating the interior of the Moon, Earth, or Mars, all come down to assumptions you have to make about what's ticking down there.

In other words, very little is definitively known about Mars. That's reflected in how geologists have divvied up martian time. Although increasingly spotty as you go further back in time, there is enough evidence to split up the past 4.6 billion years of Earth's history into eons (billion-year timescales), which are broken up into plentiful, shorter subdivisions. Mars has just four eras with no subsets whatsoever. That's it. And the first era has gone missing.

Between 4.6 billion and 4.1 billion years ago, huge impacts obliterated the surface of Mars, eradicating all the geologic evidence of the planet's earliest days. These impacts were so powerful that they sent molten and vaporized rock beyond the atmosphere. Raising the surface temperature by hundreds of degrees within an instant,[14] large bodies of water would have been annihilated. Some of that water would have rained back out over the subsequent years—thunderstorms as the world burned.

The time between 4.1 billion and 3.7 billion years ago is known as the Noachian. Big impacts still happened but were less frequent. Huge volcanism resurfaced much of the planet. Far from being a desert world, it is thought by many that plentiful water flowed and pooled across the surface. The Hesperian, 3.7 billion to 2.9 billion years ago, still featured plenty of volcanism and flowing water, but not quite as much as during the Noachian. And the Amazonian, from 2.9 billion years ago to the present, had far less prolific volcanism and far less flowing water. There's more to it than that, but that's the gist of it.

So, with such a deficit of knowledge in mind, what do we really know about Tharsis? What made it, when it was built, and how was it built, I ask Debra Needham, our planetary scientist at NASA's Marshall Space Flight Center, who also conducts detective work on the Moon.

She laughs. Mars? Pfffft. "We don't know anything!"

HUGE VOLCANOES ON EARTH often get their help from plumes, those giant blowtorches of hot, buoyant mantle material that cook up huge batches of magma in the crust. The community's best idea for what made Tharsis is a superplume—bigger or more vigorous than the plumes we have on Earth. On the basis of the rough ages of the faults, folds, cracks, lava flows, and canyons at the surface, torn up and mangled by the volcanic swell as it emerged, the construction of Tharsis probably began sometime in the Noachian, continued well into the Hesperian,[15] and was still being pieced together to an extent in the Amazonian. In other words, it took several billion years to make.

Tharsis wasn't built all at once in the same manner. Every piece of it has its own geologic history, and it's frankly a mess. "But the whole thing has been going on for the majority of the planet's existence," says Gregg.

A martian superplume could have built Tharsis in one of two ways, Needham explains. The first idea is that Tharsis is stack upon stack of lava erupted onto the surface, like hell's idea of pancakes. The second idea is that "the whole bulge is basically this big pimple on Mars," she says. The mantle melts a bit, and the partly molten bits rise and keep melting as they decompress. But all this melt might stall if it reaches a particularly thick crust or its density matches that of its surroundings. In that case, plenty of molten rock is injected into the crust, causing it to expand upward and downward, as well as left and right. Some magma erupts onto the surface as lava, but for the most part the crust gets more of the magma. To get back to our culinary metaphors, it's like an éclair. In both cases, the natural buoyancy of the superplume likely helped the entire region swell upward and outward too.

The éclair model is supported by a 2009 paper[16] co-authored by Manga, and it's based on the fact that, long ago, Mars had a planetary magnetic field. Earth has one today: the up and down swirls of iron in the liquid outer core, which sits below the plasticky mantle, generates a magnetic field that surrounds the planet. Remember how those volcanic rocks on the seafloor record the strength and direction of Earth's magnetic field when they cool below the Curie temperature? Martian volcanic rocks do the same, indicating that Mars also had a planetary magnetic field. But around 4 billion years ago, martian volcanic rocks no longer kept a record of the magnetism. Something malfunctioned deep within Mars, and the planetary magnetic field disappeared.

The bulk of Tharsis is unmagnetized. Its oldest volcanic rocks would have originally recorded the planet's magnetic field as they froze. But a load of magma intruded into this ancient crust. By reheating this old crust above the Curie temperature, their magnetic archives would have been erased. And by the time everything cooled, the planetary magnetic field would have vanished, leaving nothing left to be recorded. As such widespread demagnetization can be seen across the region, most of Tharsis was probably built by intruding, but not erupting, magma. So, Tharsis is an éclair. A giant, Lovecraftian éclair.

It's entirely unknown why Tharsis is where it is, sitting in Mars's western hemisphere just below the equator. Why did Tharsis go into overdrive while Elysium, another center of volcanism, was a dribble in comparison? Nothing currently explains that. But in a rare case where most everyone agrees with the models, what happened when Tharsis was built is both clear and mind-blowing: it tipped the entire planet over.

This is one of my all-time favorite pieces of information about the solar system. It is the sort of exhilarating thing I accost complete strangers with at parties or on the street. Every planetary scientist I've ever spoken to leans back in his or her chair when I mention this fact, and we both share a "So cool!" and "I know!" moment. It's glorious, and bonkers. It almost belongs to the world of myth, except it doesn't, because it really happened.

To be fair, planetary scientists often think that planets get tipped over. Planets tend to have an axial tilt. If a planet is spinning around on an axis that is at right angles to its journey around the Sun, it has an axial tilt of zero degrees. If it has a significant axial tilt, then maybe something knocked it over. Earth wobbles back and forth a bit, but it has a tilt of around 23 degrees. Perhaps Theia, the impactor that might have made the Moon, knocked Earth over. Uranus has an axial tilt of nearly 98 degrees, meaning it has been pushed right over onto its "back." A particularly giant impact is the prime suspect.

It's not just that the planets are knocked over, like pins at the end of a bowling alley. These impacts are so energetic that they cleave off a lot of the planet's mass into space. On Pluto—yes, it's now considered a *dwarf* planet—you have a huge crater named Sputnik Planitia. After an impact carved that out, it's thought that water from below in-filled the crater and froze into a heavy mass of ice. This huge redistribution of mass, from impact to icy in-filling, changed how Pluto spun on its axis. Imagine you have a ballerina spinning on the spot wearing three weighty rings on each arm. Now, imagine these rings change position so that there are five on one arm and one on the other. The ballerina's balance would be thrown off, meaning she would have to change how she spins or risk tumbling over. That is essentially what happened to Pluto: it tipped over in order to accommodate this big change in the distribution of its mass and to stop it from wobbling about like a one-legged penguin.

But in all these cases, giant meteorites are the instigators of the fall. Mars tipped *itself* over by creating a volcanic barbarian so huge that it redistributed its own mass. It practically turned a large sector of the planet inside-out. As that superplume put together a volcanic monster three times the area of America, it caused the planet to gradually fall backward. After a billion years or so of construction work, around the time the Hesperian was transitioning into the Amazonian, Mars had tipped over on its axis by 20 degrees.[17] If that happened on Earth, the United Kingdom would move to where the Arctic currently is, and South Africa would be just shy of the equator.

Tharsis didn't just knock Mars over. It broke it into pieces.

Closer to home, right atop Tharsis, we can see a chaotic network of crisscrossing cracks and canyons to the left of another standout feature, one great big chasm with miniature troughs splintering off from it. The former is excellently known as Noctis Labyrinthus (Maze of the Night); the latter, Valles Marineris (Mariner Valley). It's difficult to convey just how sizable these features are, but luckily Gregg is here to help. Sharing her screen with me during our call and flying over Valles Marineris, she orbits her cursor around one of the hundreds of minuscule cracks coming out of the big east-west valley. "Do you see this canyon right here?" she asks. "That is the same size as Earth's Grand Canyon." Valles Marineris as a whole is as long as the United States, is often almost twice as deep as Mount Fuji is high, and it spans roughly a fifth of the circumference of Mars. Noctis Labyrinthus is maybe half the diameter, but it's still big enough that entire volcanoes can be spotted inside its valleys. (Don't worry, it's not just you: I can't grasp these dimensions either.)

It's largely thought that both these features, the claw marks of a galactic giant, formed in part thanks to Tharsis's growth. As the crust bulged upward and outward, parts of it cracked. Noctis Labyrinthus, made of slightly different rocks and in a different location on the bulge, broke apart in multiple directions, while Valles Marineris snapped open in a largely east-west line. Both were later smoothed out and broken up a bit more by the flow of water.

Tharsis's influence didn't remain local. "Tharsis is so big that it deformed the whole planet," Manga explains. The sheer weight of it caused the region to essentially sink, warping the entire planet's crust as if it were made of plastic. And once upon a time, it may have deformed the shorelines of a northern ocean basin, filled with water that kept appearing and vanishing as if by magic.

TODAY, MARS IS A DESERT, but it was once wetter—way wetter. And this is arguably Mars's greatest mystery: How did a waterlogged planet turn into a hostile, largely uninhabitable place? Or, to put it another way: What brought about the end? Did Tharsis, being such a prominent feature of Mars, play a role? And if so, was it a hero or villain?

There is clear evidence that water on Mars still flows today, albeit in a very limited number of places. Dark streaks of soggy soil appear to ebb and flow[18] as the seasons change, indicating the presence of trickling water. The planetary science community largely agrees, though, that water flowed all over Mars far more conspicuously in the distant past. Thanks to our orbiters, landers, and rovers, scientists have spied ancient lake beds, river valleys, deltas,[19] alluvial fans, and other sedimentary features that can only be explained if water made them. Edwin Kite, a planetary scientist at the University of Chicago, points to bowls within Eridania, a region in the southern highlands that once held a large inland sea 19 times the size of Lake Michigan.

The evidence for the presence of water is not solely based on sedimentological graffiti. Orbiters armed with instruments named spectrometers are able to detect the energy signatures emitted by various chemical compounds in the martian soil. This means they can see what parts of the surface are made up of certain minerals. The *Mars Express* and the *Mars Reconnaissance Orbiter*, in 2006[20] and 2009[21] respectively, found patches of hydrated minerals all over Mars, from clays to carbonates. NASA's *Curiosity* rover, too, has found water-made minerals left behind by long-gone lakes[22] and streams[23] in Gale Crater, its martian abode.

Mars's surface is uncomfortably cold: –81 degrees Fahrenheit, on average. But all that ancient flowing water suggests it was once relatively balmy. Rivera-Hernández specializes in sediments, and she's a member of *Curiosity*'s science team. She has also conducted fieldwork in Antarctica to try and figure out what ice-deposited sediments on Mars may look like. While roving through Gale Crater, she says, they have seen no evidence for anything that was deposited by an ice-covered lake. There is no evidence that glaciers existed in the crater either. These sediments are around 3.7 billion years old, and when they were put there, the climate was warm-ish.

With all this in mind, you may gaze upon the lowlands dominating the planet's northern hemisphere and wonder if they too were once filled with water, giving Mars an expansive oceanic hat. But to suggest so to a crowd of planetary scientists would cause a ruckus—some may call you misguided, while others may enthusiastically agree. The problem is that there is ample evidence for both possibilities, and it is often unclear which lines of evidence are more convincing than others. I tell Gregg I was wading into this debate. "Oh, I wouldn't go there," she replies, purposefully shaking her head.

But how could I resist? We're talking about the end of the world here, a planetwide apocalypse through dehydration. If the Pacific Ocean suddenly up and disappeared on Earth, we would surely like to know how such a feat was possible. And whether or not there was a huge northern ocean on Mars has huge implications for the planet's climate and friendliness toward life, which, according to Earth, bloody loves water.

There is a decent amount of evidence in favor of a northern ocean. Water doesn't just come in the classic H_2O flavor. There exists a rarer, heavier version of water, one that contains a bulkier version of hydrogen named deuterium. Classic water has an easier time escaping the gravity of a planet, whereas heavy water has a bit more difficulty doing so. The natural ratio of hydrogen to deuterium is known to scientists, so the fact that there is a substantial amount of deuterium lingering in the martian atmosphere suggests to scientists that there was once a lot more classic water on Mars than we see today—enough, perhaps, for an ocean to cover up to 20 percent of the martian surface.[24]

The lowlands certainly seem like a good place to put an ocean. There are long, more or less continuous features around its edges that look, to some, like shorelines, implying there were multiple northern oceans of differing sizes and locations through time. The problem is that they aren't all flat and level, which doesn't make sense; the shorelines on Earth are flat because the seas they encircle don't disobey gravity and jump up and down at random. If you walked along Mars's shoreline-like features, you'd rise and fall several miles as you did so. But, says Manga, "Tharsis is so big that it deflects the surface of Mars by many kilometers. Something that used to be flat is no longer flat." Reversing the growth of Tharsis irons out those shorelines.

There are also giant channels,[25] dating to around 3.4 billion years old, that some have interpreted to be outflow channels for underground caches of water. When the geologic plug was pulled, catastrophic floods filled up the lowlands to make ephemeral oceans. A 75-mile crater[26] in the lowlands has also been compared to Earth's Chicxulub, the dinosaurs-ending asteroid impact scar: this martian crater is suspected to have formed when a meteorite splashed through an ocean and slammed into the seafloor, causing mega-tsunamis that crashed over the land behind the beaches, flowed uphill, and left grooves in mounds and mountains.

But all of this evidence is contested. As much as Manga buys into the idea of a northern ocean, he also tells me that we don't know if a martian shoreline *should* resemble Earth's own. Kite explains that many river deltas may not be deltas at all, and even of the few that really do resemble deltas, none of them drain into basins that could have fed a northern ocean. Needham says that those outflow channels don't all look alike. She studied how lava erodes out its own channels for her dissertation, and some features on some of those outflow channels may have been carved out by molten rock, not molten ice. Sure, salty minerals left behind by the evaporation of large bodies of standing water have been found across Mars. But as Rivera-Hernández observes, salts were not found in the northern hemisphere, which is the opposite of what you would expect if it was once home to a giant ocean.

Many argue that persistent bodies of water should only be possible if the martian atmosphere was thick enough. A substantial atmosphere with enough greenhouse gas agents, like water and carbon dioxide, could keep the surface warm enough to let the water remain liquid. It also needs to be thick enough to keep the atmospheric pressure high enough. Higher pressure prevents the formation of bubbles in liquids—boiling, in other words, as liquids turn into gases. Remove an atmosphere and even chilly water can boil off into space.

Herein lies the biggest problem with not just a northern ocean, but the presence of any lakes, rivers, and their ilk on ancient Mars: despite plenty of evidence for flowing water billions of years ago, the planet's atmosphere was stolen away in its formative years. NASA's *Mars Atmosphere and Volatile Evolution* (*MAVEN*) mission, launched in 2013, fell into orbit around the red planet and studied the evolution of its atmosphere. Bruce Jakosky, *MAVEN*'s principal investigator at the University of Colorado, Boulder, tells me that a key objective was to see whether the atmosphere was being lost to space today, how much was escaping, and why. "If we understand these processes, then we can try to extrapolate them back in time," he says.

All planets are losing their atmosphere. Earth is losing a bit today because the Sun's solar wind, its onslaught of energetic particles, is chipping little bits of it away into space. No need to panic though; it's an insignificant amount. This is partly thanks to our planet's magnetic field, which deflects a large portion of the solar wind and protects our atmosphere from simply blowing away like confetti in a hurricane.

Mars is currently losing just a few pounds of its flimsy atmosphere per second, says Jakosky. But billions of years ago, when the Sun was thought to be dimmer but pumping out potent gusts of solar wind and corrupting ultraviolet light, 10,000 times more of the martian atmosphere was being stripped away. The trapped gases within martian meteorites, when compared with present-day measurements by robots[27,28] suggest that most of Mars's early atmosphere had been obliterated just 500 million years or so after Mars was born. If true, that implies Mars was almost always an acutely frigid place with a very low atmospheric pressure. It would have been incredibly difficult, if not impossible, to have long-lived river systems, lakes, and oceans much later in its history—and yet the once wet valleys and basins suggest otherwise.

So did the planet dry up or didn't it?

Most researchers I spoke to are coming around to the idea that Mars was never warm and wet, but icy and damp. The ancient martian atmosphere would have had so little carbon dioxide that it would be impossible[29] to get the global Mars temperature above freezing and keep it there. Fortunately, that's no problem for liquid water. In Antarctica's McMurdo Dry Valleys, you still have substantial lakes trapped beneath ice. The mercury only rises above freezing for a couple of days per year, but that's enough to melt glaciers and snow to supply meltwater to underground lakes. "These persist all year round, and some of these lakes are thought to have been around for the last 1,000, 2,000 years in this region," says Rivera-Hernández. Seeing as it has water-carved features being made well into its youth, Mars may have been like Antarctica for much of its past after its original atmosphere was obliterated by the Sun.

But a northern ocean? That's a lot of water to keep in one place without it quickly freezing up and then vaporizing into space. Could the red planet really have once been a blue marble, like Earth?

IN MANY INSTANCES, I've found that the most informative scientists to talk to are not those belonging to the elderly echelons of academia, but rather graduates and postgraduates who are still working on or have recently completed their dissertations. Unlike the higher-ups, these scientists have just spent the past few years of their lives on a small handful of scientific problems, so they tend to know more about them than anyone else. They are also less entrenched in their opinions than many senior academics.

Ashley Palumbo, a recently minted doctor of planetary sciences from Brown University, spent much of her time studying Mars's early atmosphere, hydrology, climate, and geology. In 2020, she zeroed in on the paradoxes of water on an icy Mars herself. She really knows her stuff, and she says it may not be a case of whether the climate models are wrong or the geologic evidence is wrong. "In truth, it's probably that neither one is wrong," she says.

A planet's first atmosphere is often sourced from its volcanoes. They can jettison a lot of water and carbon dioxide into the atmosphere—great for global warming, as they stay in the skies for a considerably long time. Sadly, there was likely never enough of either on Mars to sustain a thick, insulating atmosphere. Other warming agents such as hydrogen and methane are very good at their job, but they linger up there for a mere 10,000 years or so. If Tharsis and its volcanic chimneys had protracted, prolific, gassy eruptions, it may have been enough to produce blips in time where things were warm enough for liquid water to be happy. But we don't have any real idea if this was the case, because the ancient magmas that contain these gases are buried miles below the surface. We can't rely on Tharsis, then, to have saved the world.

We might be thinking about this the wrong way. Earth has a "globally averaged temperature," a temperature that, on average, a point on the planet's surface will be experiencing. "But temperatures can change pretty dramatically across the surface of a planet," says Palumbo. You will burn in the Sahara but freeze in Siberia. So perhaps Mars's globally *averaged* temperature never rose above freezing, but near the heavily sunlit equator, it might poke above freezing in the warmer summer season, for example.

Equatorial or regional warming spikes, including those driven by big volcanic sneezes, still aren't enough to get a northern ocean. But, explains Palumbo, when Mars's atmosphere was somewhat thicker in the very distant Noachian past, temperature was dominantly dependent on altitude and not dictated by proximity to the chillier poles. "The whole planet kind of mimics a mountaintop," she says. "The higher elevations are colder than lower elevations." The martian highlands in the south will be cold, while the lowlands will be warmer. If the lowlands are sufficiently warm, any water trapped underground will flow onto the surface, perhaps feeding an ocean. Her own climate model results show that you only need small patches of the lowlands, rather than all of it, to remain above freezing for a millennium—the blink of an eye, on geologic timescales—in order to get northward floods of water.

This means, if the average global temperature was roughly the same as much of Antarctica, you could get a northern ocean. "This ocean would certainly be the consequence of a warmer climate, but it doesn't necessarily mean we're talking about beachballs and cabanas," says Palumbo.

There may have been a thicker atmosphere in the Noachian Era, but by the Hesperian, Mars's atmosphere was a shadow of its former self. Things were now extremely cold, even for Mars, and a bulky underground cement of ice would have prevented any liquid water buried underground from bursting through and flooding out of those outflow channels into the north. No worries, says Palumbo: models of Mars's underground temperature profiles show that, if something was able to crack open the cement, then you'd get yourself a short-lived ocean that would, in just a few centuries, be vaporized. One way to fragment this cryospheric cement would be to inject magma into or near it, which implies volcanism may have been directly responsible for any Hesperian oceans.

No one can explain where on Mars these oceans keep periodically hiding away in-between wetter spells. They appear to just vanish and reappear. But their existence is certainly possible, even on a chilly Mars. And if Mars had managed to keep a thick atmosphere, it may have always been warm enough to have rivers, lakes, and oceans. Either way, the Sun's destruction of its skies proved to be its downfall. Any life that may have existed on the ground would have struggled without the protection of a heat-trapping atmosphere, one that also shielded the surface from a constant bombardment of sterilizing solar radiation. So why does Mars lack a decent atmosphere, while Earth's still stands?

Here is where Tharsis—and all of Mars's volcanic rocks—played a key role. It was neither a hero nor a villain. It was a witness.

The magnetic minerals forged by volcanism are how we know what a planet's magnetic field was like throughout time. Strong, global magnetic fields deflect much of the Sun's hyperenergetic stream of particles—and Mars, for a time, had one of its own. But the magnetic minerals made by Tharsis and company reveal that it no longer possessed this vital shield 500 million years after it was born. Scientists aren't sure why, but it's presumed that Mars, a small world, had cooled down so much by that point that it lost the ability to churn iron about in its core. Its magnetic field shut down, leaving Mars at the mercy of the Sun's youthful rage. The atmosphere was obliterated, and a once-habitable world came to an end—a tale only knowable thanks to Mars's volcanic record keepers.

That's just one version of the story. Jakosky explains that almost every physical process that can remove atmospheric components into space depends on the behavior of the magnetic field. But we don't *really* know how much of a detrimental effect the collapse of the magnetic field would have had on Mars's early atmosphere. Befuddlingly, Mars and Earth are losing comparable amounts of their atmospheres today, even though Earth still has a magnetic field. "There's too much we still don't know about the processes operating today on Mars or Earth, let alone 4 billion years ago," he says. Unless we invent time travel, we may never truly know why Earth's atmosphere prevailed while Mars's failed.

For now, the unknowns overwhelm the knowns. But in one possible timeline, billions of years ago, a global magnetic field once presided over an atmosphere capable of keeping Mars just warm enough to allow an ocean to exist on the shores of Tharsis, back when floods of lava cascaded out of cracks in the surface. Imagine sitting on a boat, letting the alien waves carry you along as volcanic embers illuminated the night draped over the shoreline. It would have been glorious.

AS IMPOSING AS the planet-deforming, apocalypse-recording Tharsis rise is all by itself, it would be foolhardy not to talk about the magnificent magmatic mountains on top of and beside it.

Sitting a little to the northwest of the great bulge is Olympus Mons. This titan is three times higher than Everest, making it the tallest volcano in the solar system. At 370 miles across at its base, Olympus Mons can cover the entirety of Arizona. Its outer perimeter forms a cliff six miles high, meaning its edges alone are as tall as Mauna Loa, Earth's largest volcano.

Hiking Olympus Mons would be a lengthy, surreal process. Its base is so much wider than its peak is tall that you wouldn't even notice that you were walking up an incline. Right until the very last moment, you wouldn't be able to see over the peak as it would spend most of the time beyond the horizon. But what a rewarding hike it would have been a few billion years ago: there would have been all kinds of gigantic lava lakes at its summit caldera—a caldera so huge that, says Gregg, if you were to stand in the middle, the very edges of the summit would be hiding beyond view behind the curvature of the planet. And as you ascended the flanks, you may have seen waterfalls of lava splash down onto the land far below.

This all depends on whether you are lucky enough to even get your hike started, which requires that you jump over the fiery trough that surrounds Olympus Mons at its base. Being so massive, Olympus Mons has depressed the crust it is built on, "just like when you sit on a mattress," says Gregg. This made a trough around the volcano that often filled up with lava, meaning Olympus Mons once had a moat of molten rock.

Even if you managed to traverse this obstacle in a feat of breathtaking athleticism and you breached the walls of the volcanic fortress, you might be caught up in the solar system's most epic landslide. A huge rippled, rubbly deposit that radiates out as much as 450 miles away from the volcano's northwestern sector is thought to be the result of some sort of mass movement, possibly a landslide or series of cataclysmic landslides from Olympus Mons.[30]

Olympus Mons finds itself in good company. Atop the Tharsis rise itself sits a cornucopia of volcanoes, many of which are, for Mars, somewhat small. Jacob Richardson, a geologist and volcanologist at NASA's Goddard Space Flight Center in Greenbelt, Maryland, has spent several years mapping out more than a thousand of these volcanoes, and their vent sizes range from a couple of hundred feet across to a whopping 32 miles across—up there with the biggest volcanoes on Earth.

"Yeah," says Richardson. "They're still wicked large." These volcanoes were likely built by worms of magma that wriggled their way to the surface. They built a volcano and then froze, blocking the pathway to any future magmatic worms, thereby keeping these volcanoes "small."

This feeble fate was rejected by Tharsis Montes, the trio of shield volcanoes proudly rising above the rest of the rabble: Ascraeus Mons in the north, Pavonis Mons a little below, and Arsia Mons further to the south, all roughly the same age and younger than most other volcanic features on Tharsis. They probably all had their own individual, massive magma reservoirs, and partly because of this, they are humungous.

Take Arsia Mons: its summit caldera alone is 70 miles across, a crown atop a 270-mile-wide cone of solid lava. Richardson compares it to the Askja volcano in Iceland, essentially a giant hole. I've been to Askja. Peering off the caldera rim into the icy depression is an exercise in feeling insignificant. Walking inside it, Richardson imagines, is a bit like walking the Arsia Mons caldera, or at least a scale model of it. He thinks about it for a moment, then estimates that the martian caldera is 30 times larger than the Icelandic one. "Arsia Mons is . . . really large," he says, fully grasping the mind-boggling dimensions of the extraterrestrial expanse at that very moment.

The Arsia Mons caldera is spacious enough to host at least 29 individual volcanic vents. Each vent made its own lava flood equivalent to the 1783–1784 Laki eruption in Iceland—a staggering paroxysm that chucked out four cubic miles of lava, easily enough to smother the world's largest cities. And those huge outpourings were nothing compared to the truly gigantic eruptions of lava that emerged from the rift zones along these shield volcanoes' flanks: a bit like the Hawaiian shield volcano Kīlauea's rift zones, but in overdrive. One such eruption made enough lava to smother the entire United Kingdom in just a few weeks.

Some scientists take the Hawaiian comparisons to heart. Sarah

Fagents, a planetary volcanologist at the University of Hawaiʻi at Mānoa, can usually be found clambering over Hawaiian volcanoes. When she was a postdoctoral researcher in the 1990s, she joined a field course run by scientists Peter Mouginis-Mark and Scott Rowland, who used the volcanic realm on their doorstep to simulate what it would be like to trek over the red planet's volcanoes. Today, she helps run the course, the next best thing to being an astronaut on Mars herself.

"It's designed to take planetary science students . . . away from their computers and plonk them in the field to show them what these things looked like in the flesh," she says. "It's been really satisfying. You get a lot of people getting their *aha* moments when all they've done is look at a lava flow from space, and now they're there on the lava flows watching them form."

Hawaiian volcanoes like Mauna Loa or Kīlauea really are scaled-down versions of Tharsis Montes. To put it another way, Tharsis Montes is Hawaiʻi, writ large. Ludicrously large. Gregg says she has a sign in her lab that says, "Everything on Mars is bigger." The question, naturally, is why?

The lower gravity plays its part, says Fagents. With less of a gravitational crush on the crust, it's harder to close up cracks that magma sneaks through. "This means you can pump more magma up," she says. But that doesn't explain the frankly bonkers amount of magma that makes it to the surface.

It would help to understand how these volcanoes made themselves. Sure, all shield volcanoes are made up of stacks of lava that pours out of a vent in stages, but how quickly did something like Olympus Mons appear? Knowing this would allow scientists to estimate the proficiency of the magmatic forces belowground that permitted it to grow.

Lava flows at the surface of Olympus Mons have only a few craters, suggesting they are mostly around 200 million years old. Only bits of older lava flows from Olympus Mons's earliest days are exposed at the surface, but some appear to date back 3.7 billion years. Clearly, the volcano has been erupting for billions of years, but this alone can't be used to work out how *fast* Olympus Mons grew.

Lauren Jozwiak, a planetary geologist at the Johns Hopkins University Applied Physics Laboratory, made this the subject of her undergraduate dissertation. She reasoned that lava, like any fluid, flows downslope. Lava flows around Olympus Mons usually obeyed this basic rule of the universe. But several didn't, appearing instead to choose other, more circuitous and even oblique paths, including several flows in the mountain's moat. Long ago, these flows did slide downslope, but the growth of Olympus Mons and its moat kept changing the layout of the land, twisting older lava flows seemingly out of place long after they had set.

One of these so-called discordant flows was dated to be 3.7 billion years old. By winding back time and untwisting this ancient lava stream, Jozwiak and her colleagues could see where it would have been when it first flowed—and, importantly, what size and shape the volcano would have needed to be to allow that lava to flow in that exact manner.

That flow's original orientation was only possible if no more than 29 to 51 percent of Olympus Mons had already been built. Let's say 40 percent of the volcano had been constructed by the time that lava went downslope. That means this lava flow formed during the early growth stages of Olympus Mons.

But when did its construction more or less come to a halt? Those huge deposits to the volcano's northwest that may be the result of huge landslides don't appear to be covered by lava flows, suggesting they appeared after the major volcano-making eruptions had stopped. Crater counting suggests that the landslides are a little over 2.5 billion years old.

Between the emplacements of the old lava flow and the landslide debris, then, the majority of Olympus Mons—perhaps as much as 71 percent of it—appeared. That means the tallest volcano known to science was built in just over a billion years.[31]

A billion years may sound sloth-like to you and I, but we're talking about a true titan of volcanic architecture here. For most planets, this is fast. It's the equivalent of someone assembling a Jenga tower as tall as the Eiffel Tower in less than a minute. But, strangely, when you compare this feat of volcanic construction to Earth, its build speed is pretty average: making Olympus Mons in that time required a magma supply rate very similar to that involved in making the entire Hawaiian archipelago, a mantle plume–powered volcanic wonderland. So why is Olympus Mons so much bigger than Hawaiian volcanoes?

The answer, as ever, lies deep below. Mars's magnetic field wasn't the only planetary mechanism that failed. Plate tectonics, the artist that gave Earth mountains, ocean basins, explosive volcanoes, continents, and huge earthquakes, failed on Mars too.

NOTHING ON MARS suggests that it is covered in anything more than a single lithospheric lump—a slab of crust and upper mantle material, or one single tectonic plate. There is no convincing evidence that separate plates subducted into trenches on Mars, allowing them to be recycled. There are no mid-ocean ridges, no East African Rift–like zones, no mountain ranges forged by two colliding plates. "Plate tectonics did not happen on Mars," says Needham. The surface never shifted about like a jigsaw puzzle dropped into a swimming pool.

Mars's enormous volcanoes suddenly make sense. Mantle plumes, aiming at tectonic plates on Earth, have moving targets to hit. On Mars, their targets are stationary. Whether it's Tharsis powered by a superplume or Tharsis Montes and Olympus Mons fueled by their own Hawaiian-esque plumes, it matters not: if the rocks above them remain immobile, then these plumes will keep on cooking up batches of magma at the exact same spots above them for as long as they can persist.

No one knows how long mantle plumes can exist on Earth. The seafloor is recycled every 200 million years through subduction, erasing any seafloor scars indicative of older mantle-plume volcanism. Mars's mantle plumes probably got weaker over time as the planet cooled, slowing the growth of these mega-volcanoes as eons passed. But the youthful nature of their lava flows makes it clear that these plumes and their magmatic caches can live for billions of years.

It's possible that plate tectonics tried to get going shortly after Mars made its cosmic debut. NASA's *Mars Global Surveyor* recorded the remnant magnetism in the planet's crust as it orbited above, and it found an ancient region in the southern highlands with magnetic stripes. They look like the zebra stripes you get on either side of the Mid-Atlantic Ridge, but at the surface there are no features to indicate that anything close to seafloor spreading happened. It's not clear what the stripes are, says Needham, but the idea that this could be evidence of seafloor spreading isn't widely accepted within the community.

Others wonder whether the large east-west chasm of Valles Marineris is a possible false start to plate tectonics on Mars. Most agree it cracked open because of the inflation of Tharsis. Perhaps, if underlying mantle currents had succeeded in tearing it apart, you'd get a rift zone a bit like the one in East Africa.

But even if it did get started, it stopped rather abruptly just a few million years after Mars appeared, leaving the world without a meticulous planetary sculptor. Instead of having the geologic equivalent of a Michelangelo, our neighbor was left largely at the helm of a Jackson Pollock, creating chaotic scenes with the liberal deployment of meteorite impacts, splashes of water, and staggeringly effusive eruptions.

Earth is the only planet in the solar system that has plate tectonics. Knowing why is vital to understanding what makes Earth an apparent rarity. But right now, no one does.

It's not like Earth plays by its own rules. "The laws of physics work the same no matter where you are," says Needham. All that scientists can do is see what Earth has that other rocky worlds, such as Mars, lack. And Earth sure has a lot of water.

For plate tectonics to work, bits of rock need to rise up or fall below each other along faults. Water helps grease faults, from the smallest to those the size of countries. Water also makes it easier for the mantle to melt. But this doesn't just permit the creation of lots of magma. Plate tectonics describes the ability of solid but deformable rocks to tear themselves apart, says Jozwiak. The upwelling mantle wants to rip open the lithosphere, "and having water in there weakens the material just enough that it's able to pull it apart." Water also helps to maintain the presence of the asthenosphere, the weaker and more plasticky layer of the mantle that the rigid lithosphere above can sail about on in the form of tectonic plates.

Perhaps the annihilation of the martian atmosphere left the planet a little too cold, says Needham. Water would have stalled or frozen near the surface, leaving faults sorely lacking in slippery goodness and giving the planet a dry mantle unable to melt and deform.

Well, maybe. "Until we can figure out why plate tectonics started on Earth, we won't really be able to understand why it didn't start in other places," says Jozwiak. All we know for sure is Mars underwent a planetary-scale malfunction, turning it into a freakish, frigid desert world full of overzealous, oversized volcanoes.

All things considered, Mars may sound like a strange planet. But that's only because Earth is our presumption of normality. Earth, with its fancy plate tectonics, a huge diversity of volcanoes, a breathable atmosphere, vast oceans, and all kinds of living things, seems like the oddball the more you stare into space. Then again, each world has some idiosyncrasy that you can't find anywhere else. "That's the beauty of planetary science," says Jozwiak. "They're all the weird one!"

WHAT HAPPENS INSIDE PLANETS and moons dictates the fates of those worlds. And with the sole exception of Earth, the insides of every other world in the universe are largely mysterious. That is why NASA's *InSight* lander is so exciting: it's the first fully equipped geophysicist we've sent to Mars.

This robot, which arrived on Mars in November 2018, doesn't poke rocks about or shoot them with lasers. Instead, it listens for "marsquakes," gobbles up magnetic whispers, and takes the planet's temperature. Anna Horleston, a seismologist at the University of Bristol and part of the mission's science team, explains that *InSight* aims to map out the internal goings on of the planet. Scientists can use marsquakes to determine the makings of the subsurface, including the probable compositions of the mantle and core, as well as their sizes; the remnant slivers of a magnetic field will tell us how the atmosphere is socializing with starlight; the temperature of the soil indicates how much heat is escaping from below, which gives us an idea of how active or dead Mars's subterranean fires may be.

InSight, which is still gathering data, is a rarity. Robotic missions to Mars study plenty of aspects of it, but certain fields—the study of the climate, the hydrology, habitability, biosignatures, for example—get more attention than anything geophysical. Although the focus on these topics is understandable, the only way to know why Earth is a veritable paradise and Mars isn't is to study planetary viscera, to see what may or may not be cooking down there. *InSight* is a good start, but future missions need to take a closer look at Mars's volcanoes.

"Volcanism takes samples of the planet's interior throughout the history of the body and puts it on the surface for you," says Jozwiak. These "perfect slices of time" allow us to trace the evolution of the inside of a world and match it up with what was happening on the outside—you know, where everything else is.

If we ever hope to understand what happened to Mars, we "can't consider the individual components in isolation," says Jakosky. "You have to ask how they play together." Understand Tharsis, home to the most extreme volcanoes ever known, and we will fill in missing chapters in the red planet's story.

Even if you do agree that the search for signs of life takes precedence above all else, then perhaps the next sample return mission shouldn't target any of the old lake beds, river channels, or lowland plains. As we now know, Earth's volcanic environments host microbial civilizations. Tharsis, says Jozwiak, had "large outputs of heat, and you have a very rich mineral environment because of that. And during certain climate periods, the polar cap actually migrated to the flanks of the Tharsis volcanoes." This setup, featuring plenty of water and a lot of underground heat through volcanism, is very reminiscent of the conditions that gave rise to hydrothermal vents on Earth's seafloors, those rocky bastions of microbial life.

By the time you read this, NASA's *Perseverance* rover will hopefully still be gleefully wandering about in Jezero Crater, a pit once home to a lake fed from streams and from water bubbling up from below. But perhaps, says Jozwiak, "Tharsis is the place we should be looking for the possibility of life."

—

WHAT PLANETARY SCIENTIST wouldn't want to hop into a DeLorean or a TARDIS and go to see Mars's wetter, volcanic heyday? "If I had a time machine, that's the kind of thing I'd go back and see," says Needham. But we may not need to enlist the help of Doc Brown or The Doctor to see volcanoes erupt on Mars. I've made references to "young" lava flows on Mars, but in this case, I really do mean young. A 2017 paper[32] co-authored by Richardson mapped out lava flows in the caldera of Arsia Mons and counted their craters. The eruptive peak of the volcano has long passed, but these particular flows came out at around 150 million years old. Some vents may have even been active in the past 50 million years, 16 million after *Tyrannosaurus rex* bit the dust 140 million miles away. Elsewhere on Mars, "on the flanks of Elysium," says Manga, "there are places where young lava flows have erupted within the last 10 million years." Some of the youngest flows on the planet, says Gregg, appear to be around Ascraeus Mons, perhaps as "young as 10,000 years."

Crater counts may be pretty sketchy, but it's clear that Mars has been erupting for 4 billion years. That's rather odd, because Mars is tiny. "It's a very small planet that should have cooled down very quickly," says Jozwiak.

Having somewhat protracted volcanism isn't really that strange. After all, the Moon had eruptions for several billion years too. "Mercury is even smaller, and it's basically layers and layers and layers of volcanic eruptions. But it cooled down relatively quickly, and it's been dead for about 3.5 billion years, geologically speaking, more or less. So why has Mars been able to retain its heat for this long?" Jozwiak asks, before shrugging.

It all comes down to what was happening inside Mars. "If you have enough radioactive elements that are continuing to produce heat inside and you've got a capped planet—I mean, you're this one-plate planet—you're losing heat relatively slowly," says Needham. "On the flip side, when you're having these catastrophic events, and Mars clearly had some truly catastrophic volcanic events, that's how you're losing your heat. You're pumping your heat out and blowing it off the top with these volcanic eruptions. It's a very efficient heat-loss mechanism. You're killing yourself by being volcanically active."

Whatever it is, Mars's heat-keeping superpower has allowed it to erupt right up to the geological present. Some scientists even dare to think that it may have been erupting right up to the anthropological present.

InSight has picked up a bunch of marsquakes that appear to be originating from Cerberus Fossae,[33] a series of linear fissures in Elysium Planitia, a volcanic region just north of the equator. These could just be faults slipping, or high-pressure watery fluids moving about and cracking up rock. But some scientists, including Manga, wonder whether the movement of magma is causing these tremors. "It's exceptionally speculative," he says, but these seismic signals have a signature that looks a fair bit like the ones made by moving molten rock. Will this magma, if it exists, eventually erupt through the fissures onto the surface?

Richardson says that his goal is to be one of the first people to see an eruption on another planet that we didn't know for certain had active volcanoes. "I think that volcanism is totally active on Mars right now because of the *InSight* mission," he says. It may be the last gasp of martian volcanism, but it's still volcanism. And seeing as Elysium is named after an ancient Greek concept for the Good Place version of the afterlife, volcanologists can legitimately hope that heaven is a place on Mars.

VII

THE INFERNO

If you take a peek at NASA's information page on Venus, you'll see that the discovery date isn't clear because, as it notes,[1] it was "known by the Ancients." That may sound a little odd when you consider that the first identification of an entire planet in our solar system tends to be a moment of great celebration. But Venus is remarkably easy to spot, so much so that it has been seen ever since humans first started pottering around Africa hundreds of thousands of years ago. You've seen it countless times too, probably without ever realizing. Say you're stumbling home at dawn after a debauched night out or you're heading out to said debauchery as the Sun begins to set. Sometimes, you'll spot a bright light glimmering in the sky, trailing just behind the setting star or running just ahead of the sunrise. It may be known as the Evening Star in the former case or the Morning Star in the latter, but that's no star at all. That's Venus, our nearest planetary neighbor, illuminated by starlight bouncing eagerly off its atmospheric shroud.

Despite it being so close, scientists didn't know anything significant about Venus until 1962, when NASA's *Mariner 2* spacecraft[2] zipped by. It was expected to be Earth-like in some respects; telescopic work found that it was almost exactly the same size as our homeworld, and probably rocky, which is why some came to dub Venus "Earth's twin." Being nearer to the Sun, some scientists suspected it was a tropical world, a sweaty palace perhaps adorned with alien palm trees. But others, including a young Carl Sagan, zeroed in on its thick gassy envelope of carbon dioxide and dreamed up a world where global warming had become irreversible and catastrophic.

Mariner 2, flying 22,000 miles above the Venusian clouds, found that Sagan was right. The temperature of Venus's surface was estimated to be around 350 degrees Fahrenheit. Its surface, crushed under the immense weight of all that carbon dioxide, was experiencing a pressure perhaps 20 times what we have on Earth. So much for a tropical paradise.

The Soviet Union, not solely focused on trying to beat the Americans to the Moon, was also keen to demonstrate its technological might by launching its own robots at Venus. In 1967, their *Venera 4* spacecraft[3] arrived above the planet. But it didn't just whoosh by. A landing craft detached from the spaceship. It entered the atmosphere, screaming toward the surface at seven miles per second until its parachutes deployed. It began breathing in the clouds and sending information back to Mother Russia.

Soviet scientists were shocked by what *Venera 4* was telling them. The scales depicting the environmental pressure shot up so quickly to their maximum values that it looked like the readings were being made in error. The thermometer on the lander also jumped up—and kept rising and rising, surpassing the temperatures detected by *Mariner 2*. And 93 minutes after the probe began falling, it suddenly stopped transmitting data.

Things were far more severe than *Mariner 2* suggested. On the basis of the lander's final transmissions, the surface temperature was 932 degrees Fahrenheit; the atmosphere, 95 percent of which was carbon dioxide, exerted a pressure 75 times greater than Earth's. About 17 miles above the ground, after having just emerged from a suffocating cloud layer 10 miles thick, *Venera 4* was simultaneously cooked and compressed into oblivion. It was a metallic corpse long before it hit the ground.

Venera 4 didn't pass through any ordinary cloud layer. It is packed with sulfuric acid, a thick carnivorous soup that chews away at almost anything that passes through it and dissolves it away. It rains on Venus, but the corrosive rain never makes it to the surface; it's so hot down there that the rain is vaporized before it ever reaches the ground. It is hot enough to turn lead into a puddle. Liquid water, the one thing all life on Earth seems to need, simply cannot exist. There are no rivers, lakes, streams, or seas. The crater-lacking surface is a youthful wasteland of basalt, the universe's number one volcanic product. Thanks to all that carbon dioxide in the sky, the remains of *Venera 4* continue to sizzle to this day at temperatures higher than you would ever find on the dayside of Mercury, the closest planet to the Sun. The rocks feel the pressure of what we now know to be 92 times the pressure you would experience on a pleasant day on Earth's surface. It's the same sort of crushing experience you would find a mile underwater.

Venus, unlike most planets in the solar system, rotates backward, meaning the Sun rises in the west and sets in the east. And all that sunlight, refracted through the planet's atmosphere, coats the dusty surface in a grim orange haze. It's the worst place in the solar system; a robot eater; a pandemonious hellscape, the planetary embodiment of Death. If Venus is Earth's twin, it's an evil twin.

But a twin it remains: Venus is the same size as Earth; it is rocky and likely has a similar internal structure—an iron core and a mantle of plasticky hot rock below a thin crust; both worlds formed 4.6 billion years ago. Long ago, it could have been like our world today. But it appears to have been geologically corrupted beyond measure.

What the hell happened?

Paul Byrne would certainly love to know. Over the past few years, after meeting an increasing number of Venus-curious scientists, he came to realize that we know surprisingly little about our neighbor. "Particularly over the last four years I have become besotted with this world, particularly with the possibility that it was once like Earth," he says. "There's such a tragic romanticism about that, that I want to know more. That's what's driving me."

Like Earth, Venus is a volcanic place. But something went terribly wrong long ago. And finding out what went down will help us answer the most existential questions of them all: Are there countless Earths out there in the dark or are they chained to their destinies of becoming Venuses? Is Earth—the paradise—the rule, or the exception?

AFTER A FEW FALSE STARTS, the Soviet Union's success with *Venera 4* kickstarted a host of other victories in this lesser known space race. Launched in 1970, *Venera 7* made the first soft landing on Venus. Launched five years later, *Venera 9* took the very first monochrome photographs from the surface. Not willing to let the Soviets hog the stage, NASA conjured up the Pioneer Venus Project. It was a mission of two parts: an orbiter, which began dancing around Venus in late 1978, picking through the atmosphere while radar mapping the surface; and a small family of probes, which were sent screeching through the atmosphere, sending as much data back home as possible before they perished.

Of all the Venus missions, *Venera 13*, which landed in 1982, stands out for two reasons: it took the first color photographs of Venus's volcanic landscape, and it remains, to this date, the longest surviving robot[4] sent there. While snapping shots and drilling into the surface to analyze the rock, it was being squashed and sautéed. It lasted for a record-breaking 127 minutes before all communications abruptly ceased.

I ask Byrne precisely how *Venera 13* died. It is a question that has bounced around engineer-filled bars late at night, the sorts of conversations he has been privy to. The pressure, he says, isn't too bad for robots. We send submarines deep underwater and they are designed to withstand such aquatic crushes. It won't be the acid, because it doesn't rain on the scorching surface. So, unless the battery runs out first on robotic missions, what kills our mechanical explorers is the temperature. The electronics eat through their coolant, they get too hot, and they fizz out. "And I have asked people before: do these things explode? Because I like the idea of things exploding," he says. "No Paul, they say, it doesn't explode. It just stops working."

What, I wonder, would happen to a person unlucky enough to be exiled to Venus? I'm perversely thrilled that this macabre subject has been spoken about at length too. "The overall view is that the heat would kill you first," he says. "You'd probably hear yourself scream because the air is thick, but the pressure is such that [the air] would . . . fill you, very quickly." You'd probably live for a few seconds to experience it, he suggests. "It's like putting someone in an oven, but an oven underwater. Like, there's no way you're living." But don't worry, they wouldn't suffer for very long: the heat would mean that, after a few seconds of exposure, you'd lose consciousness from shock.

And in the clouds? You might be fine if you were in a hot-air balloon high enough, away from the majority of the sulfuric acid. But, he says, imagine if the balloon had a hole in it and you slowly descended into this acerbic pool. "You wouldn't want to live through the descent," he says. You would be slowly melted, eaten, and crushed.

Indeed, Venus is unequivocally horrible. But let's take some consolation in the fact that Venus is a world dominated by cool volcanoes. In 1983, radar images taken from orbit by *Venera 15* and *Venera 16* framed Venus as such with remarkable clarity, showing volcanic craters and lava flows everywhere they looked. Hints of such widespread volcanic landscaping had been found by earlier robots, but these missions gave the scientific community a visually resplendent slam-dunk.

In 1984, the Soviet Union launched *Vega 1* and *Vega 2*, two flyby missions that dropped off robotic balloons into the Venusian skies. This would be the last of their celebrated program. In 1990, a year after the Berlin Wall fell, NASA's *Magellan* orbiter, equipped with the best radar mapping system yet, arrived at Venus. It drew a picture of an infernal kingdom, 85 percent of which was covered in volcanoes and their solidified lava flows, one in which a lack of water meant that they have been left, uneroded, for eons.

A few mountains and highlands rose above the frozen fires. One, Maxwell Montes, was a little taller than Mount Everest. Tesserae—strange, mangled-up, crack-filled rocky rafts, masses of mingled clumps of light and dark—were spotted. Valleys crisscrossed parts of the planet. To make several of these features required some form of tectonics. It looked like Venus, just like Mars, was a one-plate world lacking individual tectonic plates. But as long as you have a churning mantle below, one that can pull the surface about a bit, you can still get things like mountains. Slabs of rock may be moving about a bit like pack ice on the ocean, says Byrne. When these slabs crash into each other in super slow motion, they compress and rise up into mountains. In other words, Venusian topography is made when big bits of rock smush, but we can't call it plate tectonics. Let's call it smush tectonics.

These lithic aberrations aside, Venus is a world ruled by volcanoes. And they range from Earth-like to the downright bizarre. Shield volcanoes, like those in Hawai'i, were found. Maat Mons,[5] the tallest Venusian volcano at 5.5 miles high, is covered in its own lava flows extending hundreds of miles across the surface; it is adorned with a summit caldera, like those on Mars, but unlike the red planet, there is no clear, widespread evidence for explosive volcanism on Venus. Almost everything molten appears to have been extruded like toothpaste, suppressed by the imposing, thick atmosphere.

Extruded is far too mechanical a word to describe the stunning lava creations covering Venus. Unlike the relatively homogeneous rivers of basalt you find on Earth, this planet is home to a diverse embroidery of lava flows. Novae,[6] star-like splatter patterns of lava, likely form as magma pushes up and fractures the land, with the molten rock later flowing through the myriad of splinters. Cobweb-esque sprawls, perhaps formed in a similar way, have been aptly named arachnoids. Coronae, derived from the Latin meaning "crown," are gigantic volcanic circles many hundreds of miles across filled with mountains, depressions, trenches, and a lot of lava. Volcanic domes rising from their surroundings, with round, flat tops coated in lava, are named pancake volcanoes.[7] There are sinuous rilles too, the same sort of riverbed-like remains of lava flows you can find on the Moon and Mars. One is 4,800 miles long, which is not only longer than any other rille in the solar system but lengthier than the Nile, the longest river on Earth.

But the *what* isn't as compelling as the *when* here. Missions to Venus found so many volcanoes that people began to wonder, are any still erupting today? Not today, geologically speaking, but right now, in the anthropological present? Back in 1983, the *Venera 15* and *Venera 16* images of recent-looking lava flows bolstered the idea that, as the *New York Times* reported[8] at the time, this could be a world "alive with volcanism." It was a revelation. Maybe Venus wasn't geologically dead, like Mercury, or geologically comatose, like Mars. Maybe it had a geologic pulse, just like home. If so, it would in effect make Earth a less lonely island in the vast cosmic expanse.

But not being able to see through the clouds, and not being able to keep a robot alive for more than a couple of hours on the surface, meant that scientists couldn't say for sure. They still can't. The evidence is all circumstantial, with no literal smoking gun. Many of them, though, appear to be pretty confident of the odds. "I've told people I'd eat my desk if we find it's not volcanically active," says Byrne. "Although the desk will be made of chocolate."

I'm sure that he won't have to consume his delicious desk. I would say, ladies and gentlemen of the jury, that Venus, today, is almost certainly erupting.

The first line of evidence is that, unlike Mars, Venus isn't a planetary embryo. It's a full-blown planet, like Earth. "A world that big," says Byrne, "how the f**k is it not volcanically active?" Surely it would have managed to keep lots of primordial heat, the embers from its birth, trapped inside, while hosting a wealth of radioactively decaying, heat-spewing elements. And then there's the fact that several parts of the planet seem to look a lot like Hawai'i. Well, a terrible, alternate version. But without designing a seismometer or camera or something that could survive on these hot spots without overheating themselves to death, proving whether these places contained Venusian Kīlaueas seemed to be impossible.

From 2006 until contact was lost in 2014, the *Venus Express*, a European spacecraft, orbited Venus in order to study its odd atmosphere and slices of its surface. One of its instruments was able to soak up the radiation coming off the planet, including the infrared part of the electromagnetic spectrum.

The varying textures and mineral compositions of the solid lava flows atop these apparent hot spots could be used to see how degraded the Venusian winds, heat, and pressure had made them. The more degraded the volcanic rocks, the dimmer their infrared emissions were. Young lava flows that hadn't been exposed to the elements for very long wouldn't be anywhere near as degraded and were likely to be glowing. Using that knowledge, a team of scientists in 2010 found[9] that plenty of lava on these hot spots appeared to humming in the infrared, suggesting they could be as young as 250,000 years old, perhaps far younger. Geologically speaking, that's practically yesterday.

Ten years later, another group of scientists baked a few minerals commonly found in Venusian lava, like olivine, in the lab, to see how they would degrade over there.[10] Olivine quickly transformed into iron oxide under such high temperatures, suggesting that any lava flows on Venus still containing olivine were likely extremely young. Using the *Venus Express* mission data to check, they found that lava flows thought to be 250,000 years old were rich in olivine, implying they had formed, well, yesterday,[11] this time in the human sense of the word.

Venus also appears to have discorporate belchers. There is a million times more sulfur dioxide, a common volcanic gas, in the Venusian skies compared to Earth.[12] There is, says Byrne, also a lot of sulfuric acid, which should break down over geologic timescales but appears to be a stable quantity, suggesting something—volcanic burps, perhaps—is replenishing it.

And then we have the coronae. Anna Gülcher, a doctoral student of planetary sciences at ETH Zürich in Switzerland, noticed that the 500 or so coronae on Venus all look a little different from each other. To determine why, she made 3D models[13] of the mantle plumes presumably responsible for making them. Out of the 133 Gülcher and her team looked at, they found that half of them couldn't be confidently explained by their simulated plumes. Some of the coronae, those with an elevated rim and an inner depression, were best reproduced virtually if the mantle plume there was dead. And 37 of them, those with a deep trench and an outer rise, could be computationally reproduced if a plume was still churning about below. Those 37 coronae are "still cooking underneath to various degrees, most likely," says Gülcher. An active plume doesn't mean eruptions are happening today. Perhaps they were causing eruptions a few million years ago. Which, Gülcher remind us geologically speaking, is "today."

It's frustrating to keep saying that Venus may be volcanically active on a geologic timescale. That means nothing, at least on a visceral level, to us mortals. We all want there to be alien eruptions now, because it is a thrill to think Earth isn't the only fully functioning planet in the solar system. But can we prove it?

Conclusive proof, the only thing that would definitively stop Byrne eating his desk, will come in the form of an eruption filmed in real time or a series of photographs recording the emergence of a suspicious volcanic patch that wasn't there moments before. "That would be amazing" says Gülcher. "But obviously to do that, we need to go back there."

This task won't be accomplished for some time. Unlike Mars, Venus isn't swarmed by robots. Only one spacecraft calls Venus its home: Japan's *Akatsuki* ("Dawn") orbiter, which began studying the planet's cataclysmic climate in December 2015. It's doing valuable science, but it still can't see the surface, and it won't catch an eruption in the act.

Venus is an inhospitable battleground of raw annihilative power. The exploration of its surface, and the unearthing of its secrets, has proved so difficult that it has been left in Mars's shadow. Very little is known about Venus, including its volcanism, with any certainty. Do you remember that Mars has just four geologic eras? Venus doesn't have any, at least none that scientists agree on. We have walked into a forest, fired up a single flare, and seen all we can for a few bright moments before darkness once again surrounded us—and from that glimpse we have tried to fully explain our planet's evil twin.

But it's not all hopeless. Scientists have enough data from those old groundbreaking missions that they know something terrible befell Venus. It may be a volcanic hellscape today, but it wasn't always that way. Once upon a time, Venus may have been another Earth.

MICHAEL WAY'S JOB is to model planetary atmospheres at NASA's Goddard Institute for Space Studies in New York. In other words, he tries to find out how planets terraform themselves.

Take Earth. It wasn't always like it is now. We know that, billions of years ago, there was a time that the entire thing would have looked like a slushy snowball. There were times that it was remarkably hot and wet, and others where it was cold and dry. There were times the sky was orange and oxygen wasn't free to be inhaled. All of these Earths, if they existed as viewable copies elsewhere in the cosmos, would look like completely alien worlds. Michael Way looks at planets as they are today and uses everything we know about worlds to fiddle with the dial of time, taking us into the past and far into the future, to see what was and what is yet to pass. His job, then, is to find out why planets become delightful or dreadful. That, luckily for us, includes Venus—when, how, and why it became dreadful. If we get reasonable answers, then we can compare Venus's saga with Earth's in an attempt to find out why these planets were prescribed divergent destinies.

Venus may have always been a lava-strewn cataclysm. But orbital missions sniffing its skies have found the chemical ghosts of water. Specifically, they found a lot of heavy water, that rarer version of classic, lighter water also found in the martian atmosphere. And all that heavy water suggests that there was once a heck of a lot of classic water on Venus, a shallow ocean's worth in fact. But when something turned Venus into an acidic pressure cooker, all that water was obliterated. What was that something? How long did the water stick around for, and in what form?

Way takes me back in time 4.6 billion years, right back to Venus's formation. Before he begins his story, he warns me that, like Camelot in *Monty Python*, he and his colleagues' work[14] is only a model. Nothing is known for sure. But, with that in mind, he begins the game, a Choose Your Own Misadventure where the fate of Venus will be decided. Many paths will bring a pleasant Venus to an end. Only one will keep it paradisiacal. Will we reach the promised land?

At the start of the game, Venus, like all baby worlds, was probably covered in a magma ocean as it orbited a newly born Sun. Being closer to the Sun than Earth, it would be a bit warmer by default. It may not be possible on some worlds that close to their run-of-the-mill star to keep any water from simply boiling off into space. But a day on Venus is painfully long, so long in fact that the clouds blanketing the dayside would just linger there, like giant vaporous spaceships, shielding the surface from the Sun's harsh rays and keeping water liquid. Some of that water may have been delivered by comets or soggy asteroids. Some of it would have erupted out of the magma ocean, seeing as water vapor is the most abundant volcanic gas.

As the Sun goes from its youth into its adolescence, it gets hotter. Let's turn the dial up, make the Sun get hotter, and see what happens. That turns all our water into unpleasant steam. Being a greenhouse gas, all that water vapor then turns Venus into a sadistic sauna. Oops. While doing that, we forgot to check on the magma ocean. Oh no! We left it molten for too long, and it released lots of carbon dioxide, a very common volcanic gas. The carbon dioxide and the water vapor prove too much for Venus; global warming keeps going and going until Venus turns into a scorched ruin. Game over.

Fortunately, we have a few lives left to continue the game.

Let's go back to the start, with a bubbling magma ocean. This time we'll tweak it so it cools and freezes over quicker, keeping much of that carbon dioxide trapped in the planet. The carbon dioxide that already escaped is dissolving in the liquid water, which stops the climate getting too hot. Fantastic! The liquid water pools as an ancient ocean, washing against the volcanic shores warmed by the filtered light of a young star. Oh, but we forgot about the Sun. It gets too hot too quickly again, and any water on the Venusian surface vaporizes into steam, warming the world. With no sink left for the carbon dioxide to fall into, it goes back into the atmosphere, and we get runaway global warming again. Venus is, once again, ruined.

Some advice from Byrne pops up. When you get to the point where you boil off your oceans, he says, "it's goodnight Gracie, game over!"

Way interjects. When it comes to the nefarious nature of the Sun, he says, there is no escaping the fact that it gets hotter as it ages. That's just physics, and we can't break the universe to suit our needs. There are no cheat codes in this game. But perhaps the difficulty setting is too high. The Sun will always bake Venus, but its doomsday effect, boiling all the planet's liquid water to steam, would only happen very early on in the game or not at all. So let's take it down a notch. The Sun will get hotter over time, but gradually—and liquid water is allowed to exist and persist.

Let's continue. Our magma ocean freezes sharpish, the clouds shield Venus from the Sun, and liquid water pools on the surface. It's tempting to fiddle with the controls a bit here, but if we leave it, Venus and its water get on fine. In fact, there is a rather beautiful ocean developing. The climate is tropical, a little warmer than Earth, on average, but doable. Millions of years pass. Many millions. Actually, this may be a pretty good place for life to . . .

An alarm goes off. The global temperature is rising.

Way pops up again. "If you don't have plate tectonics, all bets are off," he says. "If you don't have volatile cycling, you're screwed." We quickly mash at the controller and find the option to initiate plate tectonics. The planet's crust starts to succumb to the cycling mantle below. Bits of crust smash into each other. Other bits break up as rift zones form. Tectonic plates meet others and go beneath them, falling into the mantle. To our relief, we see atmospheric carbon dioxide dissolving in the ocean and turning into carbon-bearing rocks. Those rocks get dragged down into the mantle with doomed tectonic plates, keeping the carbon there for eons and only very slowly releasing it through volcanic eruptions.

Close call. It turns out that plate tectonics is a great way to regulate a planet's temperature. It's not perfect, and changes to the orbit of a planet or the chemistry of the atmosphere can cause wild swings. But plate tectonics seems to stop Venus from becoming anything irreversibly awful, at least on geologic timescales.

Millions of years pass. Things look a bit warm, but decent. You start to wonder if any fish would enjoy swimming in that ocean over there when another alarm goes off: a volcanic eruption has begun. No big deal, right? Volcanoes erupt all the time on Venus. You rotate the globe to see if you can find the eruption. You expect to find a small bleb of lava somewhere. Instead, you find a huge outpouring of lava. A whole region on Venus looks like someone spilled a god-sized amount of molten rock.

The eruption isn't coming from one volcano, but from a series of cracks in the ground. Thousands of years pass. Tens, hundreds of thousands. Lava is still erupting. Is this a glitch? Is there an off button? While you scramble to find one, another alarm goes off: a second colossal eruption has begun elsewhere on Venus. You begin to sweat. You clock the planetary thermometer. It's rising. Those protracted eruptions are releasing a great deal of carbon dioxide. What is going on? Your oceans and seas begin to evaporate. The increased amount of water vapor in the atmosphere warms the planet further. But just before you think the critical point has been reached, the eruptions stop. Carbon dioxide is no longer being pumped into the air. Is it over? Did we make it through to the other side?

Nope. It's already too late. The amount of carbon dioxide was just enough to set the ball rolling. It's far too hot, and the oceans and seas continue to vaporize. We hope that those subducting plates, the ones falling into the mantle, will rob the sky of enough carbon to save us—but something's wrong. Without any liquid water, tectonic plates can't break up properly, can't flow properly, and they can't melt as easily. The engine of plate tectonics seizes up. No more plates subduct. All that carbon isn't getting eaten up by rocks anymore.

With no huge aquatic sink left to dissolve that carbon dioxide, and no plate tectonics burying the carbon in the mantle, it stays up there. Those two greenhouse gases work together to raise the temperature of the planet to a terrifying degree. The water vapor in the atmosphere doesn't even get a chance to rain back out onto the surface. The solar radiation snaps the oxygen away from the hydrogen; being incredibly light, the hydrogen escapes into outer space. Your water is gone. Venus underwent the apocalypse. Game over.

"The way you throw a planet into a runaway greenhouse effect if it's not by a brightening Sun is you dump a boatload of CO_2 into the atmosphere very quickly," says Byrne, summarizing my failed playthrough. "And the most effective way we have of doing that is volcanism."

This sort of epic, prolonged volcanic crisis may seem like an unfair surprise. And it is. But that doesn't make it unrealistic, because at points in Earth's vast history, similar eruptions have happened.

The highlight (or lowlight, depending on your point of view) of science journalist Peter Brannen's masterwork on the mass extinctions that have befallen Earth, *The Ends of the World*, is the story of the end-Permian mass extinction, an avalanche of death that rocked the world 252 million years ago. Most living things perished. Biology barely hung on; a little more death and, microbes aside, life on Earth may have been entirely exterminated.

The prime suspect? Siberia. Around that time, a vast volume of magma was injected into the Tunguska sedimentary basin, a rocky bowl filled with carbonates, shales, coals, oil, and natural gas. Like a match to gasoline, the whole thing explodes, over and over again, scarring the planet and unleashing eye-watering amounts of carbon dioxide and methane, a more potent but shorter-lived greenhouse gas, into the sky. The eruption itself also released a lot of carbon dioxide as it continued to pour lava over Siberia for, say, 2 million years or so.[15] I can't imagine this field of fire, this continental-scale paroxysm existing for so long, but it helps me put it into perspective when I realize that this eruption went on for longer than humanity has existed.

By the time all is said and done, 3 million square miles is covered in frozen lava, a formation named the Siberian Traps. This was nowhere near enough to cover the whole world, but the greenhouse gases gushing out of that spot wrecked the climate. Anywhere from 10,000 billion to 48,000 billion tons of carbon, free at last, lights an atmospheric blast furnace. The land gets around 30 degrees Fahrenheit hotter and stays that way for an age. A similar scorching happens to the world's life-filled tropical oceans. Life was poached to death. Sure, much carbon dioxide dissolved in the oceans, but when it does so it initially becomes carbonic acid. A little bit isn't a problem, but all that carbon turned the oceans into a corrosive broth, filled with dissolving creatures. Volcanoes spend most of their time not slaughtering things. But it would be remiss to avoid the fact that, every so often, on timescales of hundreds of millions of years, they really do a number on the planet.

There is no oil or coal on Venus. But we know that all that carbon dioxide had to have gotten into the sky and baked the planet somehow. Way's models can't say how many Siberian Traps–style eruptions it would have taken to cause the oceans to begin to not just heat up but actually boil away. But there must have been many moderately sized ones or just a few mortifyingly gargantuan eruptions, because Venus didn't just get hot and then recover. It got extremely hot and never went back. The temperature rose by something like 700 to 800 degrees Fahrenheit. Its volcanism broke the world.

Time passes. Venus sizzles. With no more plate tectonics, the planet's internal heat isn't able to escape—as in, erupt out as lava—in a controlled manner, like it does on Earth along its mostly underwater ridges. Instead, the heat comes out everywhere.

Suzanne Smrekar, a planetary geophysicist at NASA's Jet Propulsion Laboratory in Pasadena, California, tells me that there are, if we simplify, two ways in which this can happen. Eruptions of lava happen everywhere, all at pretty much the same time. The surface of Venus is wiped clean, sterilized by fresh lava. Alternatively, you get a steady stream of eruptions happening in various regions of the planet over many tens of millions or hundreds of millions of years. Ultimately, this would cover the planet's surface too.

It's not clear which is correct, but the end result is the same: we get the Venus we see today. With so few impact craters, the surface, mostly made of lava, is clearly very new, anywhere from 750 million years old to 180 million years young.

Let's say the surface is 500 million years old. That means the planet's surface was almost entirely smothered in lava, as if someone knocked over a galactic cheese fondue, at some point before then. The oxygen, segregated from its space-bound hydrogen friends back when it used to be water vapor, is too heavy to zip off into the unknown. Instead, it gets soaked up by that fresh lava, where it remains imprisoned today.

The truth is, that game is rigged to fail. Venus, today, is awful. But what playing the game has taught us is that, for a brief moment, or for billions of years, Venus may have had an ocean. There may have been streams, waterfalls, thunderstorms, and lakes. It may have been a beautiful Garden of Eden right next door. It may have even been habitable—not necessarily inhabited, but amenable for life. As Byrne playfully suggests, there may have been trees. There may have been alien squirrels. We'll probably never know, because the world burned down.

The idea that Venus may have once been so thoroughly habitable is a jarring thought. But now that it isn't, we are faced with a profound question. Why did Venus go down in flames, but not Earth? The formation of the Siberian Traps was a disaster, but it didn't irreversibly corrupt the climate. Earth has had a handful of other eruptions like it, other continental-scale effusions of lava that release horrific amounts of greenhouse gases and take epochs to end, in the past few billion years. But why hasn't Earth ever had multiple eruptions like it simultaneously, as Way suspects happened on Venus?

In other words, which is the standard story for rocky planets the size of Earth and Venus? Did Venus get unlucky and suffer from a rare confluence of multiple mega-eruptions? Or did Earth get lucky and, so far, has only ever had one mega-eruption at once? Unless we go back to Venus, study it closely, and find out what really happened all those eons ago, we will never get an answer to these existential queries. All scientists can do is use the limited data they have to set up their virtual worlds, plug in their controllers, and play the game, over and over again, hoping to find clues in digital incandescence.

VENUS MAY HAVE BEEN a comfortable place for alien life in the past. Today, though, with its acidic skies and crushing, sweltering surface, most scientists consider it to be as habitable as the business end of a flamethrower.

But then, on September 14, 2020, something really weird happened: a chemical compound named phosphine—one part phosphorus, three parts hydrogen—was detected in the clouds of Venus by two telescopes on Earth. And the world went bananas.

I live in a part of London named Greenwich. It is like a little village just south of the River Thames, with breathable air, a huge historical park, and a quick train ride to the city center. It's bloody beautiful, and my Shollie dog, Lola, loves being taken for walks there. She especially loves the park, and not just because it is a veritable adventure land filled with friendly people and their own pups, but also because of the picturesque panoramas. The best view is from next to the venerable Royal Observatory, perhaps most famous to tourists for marking a now slightly mispositioned prime meridian, where the longitude is zero degrees, the Imperial-era arbitrary middle line running from top to bottom through the world.

It is Friday, September 18, 2020, and I call up Emily Drabek-Maunder, an astrophysicist and the senior manager of public astronomy at the Royal Observatory. She had already done countless interviews and livestreamed talks, Q&As, and more over the past five days. I ask how she is. "I'm absolutely knackered!" she tells me. I almost feel bad for adding another interview to the pile.

Drabek-Maunder is part of an international team of scientists led by Jane Greaves, an astronomer at Cardiff University. Her team was trying to use ground-based telescopes to search for signs of life in the solar system and trying to better understand how solar systems are made. As part of this grand undertaking, they hoped to see if they could use telescopes to look for important gases that are involved in either research topic. While looking at gases on icy moons, like Jupiter's Europa or Saturn's Enceladus, they kept coming across a problem: they were finding gases that could be made by life but could very easily be made by non-biological processes too.

Take oxygen. Way mentions to me that the presence of a lot of oxygen gas in the atmosphere, like Earth's, has been suggested by some astronomers to be a possible biosignature, a sign of life—a world full of photosynthesizing plants or bacteria, for example. But as models like his show, all that free oxygen may be the remnants of liquid water that was obliterated long ago, with the hydrogen in that water vapor escaping into space.

Ideally, Greaves and everyone else wanted to find a gas that was known to be made by life, but that was incredibly difficult to make using geology or environmental chemistry. That's when Drabek-Maunder throws me a real curveball: "We didn't know what to search for until Jane came across a paper, randomly enough, about penguin poop," she says. It turns out that penguin poop is filled with microbes, and some of those microbes produce a lot of a gas named phosphine.

Phosphine is hard to make abiotically. "You have to have a lot of energy to make it, to get phosphorus and hydrogen to kind of stick to one another because they want to be paired with other things—mainly oxygen," Drabek-Maunder explains. But life sure loves making it on Earth. Although made by microbes that don't require oxygen to live and grow, it isn't exactly clear how they manufacture it. But whether it's an accidental by-product of their biochemical cookery or they make it directly, phosphine can be found in all sorts of gloopy places, from the penguin poop–covered wilds of Antarctica to the sewers and marshes of Florida. Humans make it too:[16] it was used as a chemical weapon in the First World War and can crop up in meth labs. Walter White used it to kill someone in the very first episode of *Breaking Bad*.

Another member of the team, Clara Sousa-Silva, a quantum astrochemist at the Massachusetts Institute of Technology, had spent several lonely years studying phosphine on her advisor's advice, after her own suggestions were rejected.[17] After extensive examination, she wondered if it would make a good biosignature gas to look for in the skies of other worlds. In a 2019 pre-print paper,[18] she and others outlined this possibility, penguin poop and all, hypothesizing that phosphine could happily accumulate to detectable levels in a planet's atmosphere so long as it wasn't broken down too quickly by the Sun's ultraviolet radiation.

As a test of their ability to spot phosphine via telescope, they thought they would point one at Venus. Why not? It's an easily observable world. In 2017, after being given the go-ahead by the director of the James Clerk Maxwell Telescope sitting 13,400 feet high atop Hawai'i's Mauna Kea volcano, they had a peek.

Venus bounced back so much sunlight that it created bright reflections inside the telescope, so the data took a while to process. As work on the data progressed, Drabek-Maunder, who was a full-time researcher back then, left her job to work in public engagement and science communication. A couple of years passed. Finally, in January 2019, after the noise was painstakingly removed, the data suggested the light swimming through the clouds of Venus had the spectral fingerprints of phosphine.

The team couldn't quite believe it, because they never expected to find it. Venus was meant to be a fun and easy calibration exercise. Instead, they smelled the scent of phosphine. To be sure it had been detected, the team applied for time on a more sensitive telescope, the Atacama Large Millimeter/submillimeter Array in Chile, in March of that year. A few months later, as spring became summer, Greaves called Drabek-Maunder and showed her the readouts: phosphine was definitely flittering around Venus.

There it was, a complete shock but clear to see, like diamonds falling from the sky. "Out of every billion molecules in the atmosphere, only 20 are phosphine, but that's still an incredible amount," says Drabek-Maunder.

The team still had to write up a scientific paper, and until its scientific peer review and publication, they couldn't tell a soul. Officially, until September 14, 2020, when their discovery paper[19] became public, they had to keep mum. But we're all human. Surely, she told someone. "I told my husband!" she says. "He's known about it the entire time." Greaves, she says, told her mom.

Found somewhat high up in the Venusian clouds, this phosphine is constantly being broken down by ultraviolet radiation. Something, then, is replenishing it all the time—but what? Phosphine can be made in the hearts of gas giants, like Jupiter and Saturn, but only because of the intense heat and pressures found deep below the swirling clouds of these worlds. To make it this way on Venus, for all its impressively hellish confines, isn't possible.[20] Meteorite impacts and lightning could make it, but not in the quantities observed. Volcanic eruptions could jettison it into the sky, but for that to sufficiently explain the data, Venus would have to be 200 times more volcanically active than Earth is now. It isn't.

It was thought possible, then, that this gas is the halitosis of alien life levitating in the clouds of Venus. Forget fiction like *Contact* or *Independence Day*; our first encounter with extraterrestrials may not be through mysterious bleeps or giant motherships, but instead with stinky microbes—not on some far-flung oasis world, but on the volcanically trashed planet right next door.

SCIENTISTS BEGAN TO PONDER on the existence of Venusian life in the 1960s, a few years before heat-loving microbes inside Yellowstone's hot springs had been discovered, and long before hydrothermal vents had been found. Carl Sagan and Harold Morowitz, a biophysicist who searched for the origins of life, wrote a paper in 1967[21] where they let their imagination run free.

Sure, the Venusian clouds may be an acidic dystopia, but, they noted, there's a decent amount of carbon dioxide, water, and sunlight up there, enough to support some sort of photosynthetic life. It may be too hot low down, and too cold high up, but in the middle, it's just right. "The conditions in the lower clouds of Venus resemble those on Earth more than any other extraterrestrial environment now known," they wrote. "If small amounts of minerals are stirred up to the clouds from the surface, it is by no means difficult to imagine an indigenous biology in the clouds of Venus."

But what would life there look like? Horowitz and Sagan wondered if something up there could remain up there in the temperate zone by gathering or making hydrogen and stuffing it into a sort of biological balloon, one that collects water from moisture or rain. "The organism is essentially a spherical hydrogen gasbag," they speculated. With all that phosphine, perhaps life takes the form of the Venusian equivalent of a flying flatulent penguin.

In all likelihood, if there is life, it is probably going to be microbial. And the idea that the clouds of Venus aren't actually as terrible as they first appear, at least for microorganisms, has come up a few times in the past half-century. A 2018 study[22] noted that you get Earth-like surface temperatures and pressures in the lower cloud decks of Venus. The oceans may have long gone, but there is plenty of water vapor still up there—not all of it was split apart by ultraviolet radiation. But that acid is surely a problem, though, no? Oh yes. "Venus's clouds are nearly 100 percent acid," says Drabek-Maunder. "It's insane. The sulfuric acid will just kind of melt anything."

Kennda Lynch, our astrobiologist and geomicrobiologist from the Lunar and Planetary Institute in Houston, Texas, reminds me that in Ethiopia's Danakil Depression, "we have pHs of zero there—we're talking negative pHs there." The Rio Tinto river system in southern Spain, where you have acid rock drainage, also has negative pHs. The clouds of Venus aren't quite the same though. The acid isn't just in a big pool you can swim in if you're feeling masochistic; it's in droplets suspended in the air, so not everywhere has as much acid as anywhere else. And the interaction of ultraviolet light with the chemical cocktail up there is pretty different from the places we have on Earth, perhaps creating some "kickass chemistry we don't know about," says Lynch. Maybe, she wonders, the closest analog we have are volcanic fumaroles, belching superheated acidic gases into the air. After all, microbial life on Earth, the sort armed with appropriate heat and acid shielding, find fumaroles a nice place to live.[23]

It isn't actually that difficult for life to protect itself from acid. "William Bains, one of the chemists from the MIT team, he was doing some really great experiments, showing people what sulfuric acid does. He was literally pouring sulfuric acid on different materials and showing how it melted them and everything," says Drabek-Maunder. "But he poured sulfuric acid on a succulent and it didn't really do anything. That's because succulents have a waxy coating on the outside of them that can protect them. So it could be that if there is life in Venus's clouds it could have a shell of some kind or it could exist in droplets." Life could even exist in droplets of hydrochloric acid, shielded from the worst of the sulfuric acid up there.

We just can't assume that the genetic makeup of Venusian life is the same as Earth's. "Life could work in ways we don't think about," says Lynch. How they organize their genetic material, what their genetic material is based off, what their genetic structure is—life on Venus could have evolved in such a way that floating acid just doesn't matter.

Okay, so microbial life can protect itself from dissolution. And maybe it uses photosynthesis to grab food or maybe it eats some of the volcanic minerals being swept into the sky by updrafts. But how does it stay floating up there and stop itself from falling into the frying pan below? One answer, from a 2004 paper,[24] is that it doesn't. Sure, the microbes stay up there for some time, but eventually they will fall into the lower haze layer and burn up. The only way they would survive is if they were able to reproduce fast enough to keep replenishing their population. (This sounds terrible. It'd be like living in a house where, after a set amount of time, a trapdoor to hell opens up. Haven't had kids yet? Too bad, you're being incinerated.)

Fortunately, in August 2020, a mixture of scientists, including some of those on the September phosphine discovery team, perhaps not coincidentally published a paper[25] that proposed a less crappy living arrangement for these hypothetical microbes. On Earth, microbes don't just stick to the ground. Some have been found drifting in the clouds, lifted there by the wind. The problem is that they need to quickly get back down again because if they don't, they will dry out and ultimately perish. That suggests the only way for Venusian life to live is to live up there in watery droplets in the sky.

That's perfectly possible. It's also fair to suggest that they may have pigments that act as a sunscreen against harsh ultraviolet radiation. And life has plenty of ways to adapt to acidic conditions, so that's not much of a concern. The problem is the same as before. The microbes may act as nuclei for clouds to condense around. Eventually, individual water droplets will coalesce. And when the water droplets get too big and heavy, they will fall to the lower atmospheric haze where they will dry out. But, says the team, these desiccated spores may not all die. Many could drift about above the surface, dried up but in a suspended state of animation. When an updraft brings them into the clouds again, they encounter water and reawaken. In other words, instead of a trap door of death, they could exist on an infinite rollercoaster of near-death experiences.

Horowitz and Sagan also thought about how life would have gotten into the sky in the first place. Long before evidence of ancient Venusian oceans was found, they wondered whether life arose under more moderate conditions on the surface of Venus in its early history, and when the surface became too hot and dry, life "may have then emigrated to the clouds." The idea that Venusian cloud life could be refugees from the stellar or volcanic apocalypse that consigned the planet to its doom is an alluring notion. It isn't clear how it would have survived either event, but Earth reminds us how unyielding life can be.

We don't know exactly where life began. "But the fact is, life has had 4 billion years to fill every single ecological niche on this planet," says Way. "No matter where we look, we find it. And that's why, if there are really microbes in the atmosphere of Venus and they've found a way to survive, if it really is indigenous to the planet, it has to come from a long period of habitability." We don't know how long Venus possessed an ocean's worth of water, if it ever did. But if life got into the clouds, it would have needed a lengthy amount of time to be able to stumble upon its own Wright Brothers moment.

If Venusian life today isn't the product of an exodus, it could be the descendants of exiles. It's less likely, Byrne reckons, but still possible, that an asteroid impact on Earth sent life-containing rocks tumbling down the solar system's gravity well all the way to Venus. That would make Venusian life our very, very, very distant cousins.

Either possibility, if found to be true, would be the single greatest discovery in scientific history. If that phosphine is biological, then we know something truly remarkable about the existence of life. Shortly after the discovery was announced, Byrne spotted an amusing but profound take on it on Twitter: "Venus is the NYC of habitable environments," the tweet said. "If it can make it there, it can make it anywhere." If it can withstand those temperatures, pressures, and acidities, and it can outlast the most extreme climate change scenario known to science, life really can make it through *anything*, on almost any planet in the cosmos.

But we don't know it's life. The authors of the phosphine paper can't find a better explanation, but they don't claim it as clear evidence of alien microbes. It is, however, the most compelling evidence we have for alien life yet, says Byrne. It's possible that the methane gas found in the skies of Mars is biological too, seeing as certain microbes on Earth make plenty of it. But methane is made in huge quantities by geologic processes too, including volcanism, which may still be bubbling away on Mars today. Phosphine, in the high quantity seen in Venus's clouds, cannot be explained by any chemical or geologic processes of which we know. Unfortunately, most of Venus's chemistry and geology is opaque to us. "For me, it's basically telling us how much we don't know about Venus," says Noam Izenberg, a planetary geologist at the Johns Hopkins University Applied Physics Laboratory and deputy chair of NASA's Venus Exploration Analysis Group. We don't know all of the ways phosphine can be made, and we don't know all the ways it can be destroyed or trapped.

"Life is a likely possibility, so it's in the list, definitely," says Lynch. "But there's a lot of other possibilities we have to rule out before we can just go, yep, it's life! You know, extraordinary claims require extraordinary results. So we have to really vet this early, preliminary result."

Don't get your hopes up too much. "Occam's razor says it's not life. And I'm fine with that," says Byrne. But needing to know is what matters. "I'm hoping we'll look back at this as a watershed moment, where the conversation seriously transitions away from places where we conventionally think are safe and nice, with nice climates and water, to places that might not otherwise have ever been places we'd seriously consider containing life."

IN OCTOBER 2020, members of the phosphine team were looking at other ways to confirm their detection. By that point, some scientists had already voiced skepticism that the phosphine signal as "seen" by those two telescopes was genuine. The phosphine detection could be a phantom signal from the data, some suggested, perhaps an inadvertent consequence of all that data processing designed to remove the noise. Some were unable to find the phosphine fingerprint using other methods. A few different preliminary assessments of the discovery data itself found no clear signal for phosphine.[26,27]

But by the end of the month, a reexamination of data from the Pioneer missions unearthed a phosphine-like blip,[28] first picked up way back in the 1970s. And in November, the Greaves et al. discovery team took a look at newly recalibrated data from the telescope in Chile and found a weaker signal[29] for phosphine—but a signal nevertheless.

The scientific debates will continue for some time. Perhaps the phosphine detection is real. But it's possible that it's a false positive. If this turns out to be the case, the official nixing of the discovery would be disappointing, but this is precisely how science is supposed to work—as a perpetually self-correcting machine. That's the only way we know what is true, and what is not.

But, phosphine or no, the clouds of our planetary twin remain a potential habitat for life. That possibility will linger no matter what drama occurs on Earth.

So how will we ever know whether there is life in the skies of Venus? How will we know whether Venus was brought down by a volcanic apocalypse? How will we know whether volcanoes are erupting on Venus today? The answers to all of these questions, questions inextricably linked to one another, are the same: we need to go back.

NASA's mission selection process is drawn out and grueling. Many hearts have been broken, especially if you like Venus. From the cheaper, less technically capable Discovery-class missions to the more expensive New Frontiers–class missions, right up to the multibillion-dollar Flagship-class missions, Mars features a lot.[30] Venus, as well as Uranus and Neptune, haven't seen a single win for decades. The last NASA mission to Venus was *Magellan*, launched way back in 1989. The last European mission was 2005's *Venus Express*. Today, just Japan's *Akatsuki* spacecraft keeps Venus company. But times may finally be changing.

Stephen Kane, a planetary scientist at the University of California, Riverside, has studied exoplanets, worlds outside our own solar system, for 25 years. In the past decade, working with the exoplanet-hunting Kepler space telescope, he was present as a huge family of Earth-sized rocky worlds was discovered. "And I thought, hang on a minute," he says. "If we're talking about Earth-sized, that's the same as Venus-sized."

Unlike the scientists who sniffed out the Venusian phosphine, exoplanet hunters don't yet have the telescopic ability to peer into the atmospheres of these distant worlds. They don't know whether they are packed full of carbon dioxide or they are more like the nitrogen–oxygen mixture we have on Earth.

Knowing the difference really matters. "When people say Earth-like world, everybody's thinking Endor, right?" says Izenberg. (He means the Forest Moon of Endor, the one from *Return of the Jedi* with the trees, Ewoks, speeder bikes, and the Empire's shield generator. Endor is the gas giant that said forest moon orbits.) "Ooh, forests and jungles and oceans. But no, it's probably sulfuric hellscapes."

Sure, these exoplanets may exist in the habitable zone, the perfect Goldilocks-esque distance from the star where things are not too hot, and not too cold, so liquid water is able to exist. But how do we know whether several simultaneous mega-eruptions haven't cooked them? Volcanoes, says Way, "decide the worlds they belong to." Sometimes, like on Earth, they provide the first draft of an atmosphere, one rich in carbon dioxide that photosynthesizing plants and bacteria can use to make oxygen. Sometimes, like on Venus, they might overdo things a bit. "One thing that is pretty clear is that there are planets freaking everywhere," says Izenberg. "The question is, how everywhere are the planets like ours?"

When he began inviting himself to Venus science meetings, Kane says, "one of the first things I discovered was that they are an extremely data-starved community that look with great envy at their martian colleagues who just seem to get rovers and everything just thrown at them." But in the past few years, there has been a shift in favor of Venus. Along with several others, Kane has been arguing to NASA headquarters that if we want to understand planetary habitability, both inside and outside our cosmic backwater, we need to look at Venus, "because that's a place where it went wrong." By studying it, we will get clues as to whether there are more Earths or more Venuses in the galaxy.

It was this sort of notion that had recently begun to convince an increasing number of higher-ups of the value of returning to Venus. But now, all of a sudden, national pride has become a factor. The same day that the phosphine discovery was announced, Jim Bridenstine, the NASA administrator, tweeted that Venus must be a future priority. And then, right afterward, the head of Russia's government space agency, Roscosmos, pointed to the Soviet Union's technological legacy and claimed Venus was a Russian planet.[31] "They'll be damned if they're gonna see the Americans go there and confirm life!" says Kane.

Nationalism is a particularly pernicious poison at the best of times. But this surge of interest thanks to the phosphine is no bad thing. So much science has been funded to investigate the possibility of life on Mars, missions sometimes driven by its mysterious methane emissions. "We'd better get something similar for Venus," says Byrne.

In 1996, scientists looking at ALH 84001, a martian meteorite found 12 years earlier in Antarctica, thought they saw microscopic fossils[32] between grains, which meant there could be life on Mars! Ultimately, it turns out that this wasn't all it was cracked up to be. But this case of mistaken identity was enough to kickstart a huge dedicated mission program to Mars. This meteorite, says Izenberg, "didn't find life on Mars. But it really pushed the needle. It really gave everybody this push." Phosphine could be the equivalent moment for Venus. "It doesn't matter whether it is correct or not, because we'll find out by going there, and a whole bunch of other cool stuff."

Way before phosphine became fashionable, NASA was already preparing to decide which of its economical Discovery-class missions it was going to pick next—up to two could be chosen. At the time of writing, in the autumn of 2020, four proposals headed by different teams of scientists are currently in the running: one would go to a moon of Jupiter, one would fly to a Neptunian moon, and two of them would head to Venus. Both teams were, quite understandably, thrilled about all the attention Venus was suddenly getting.

One, the *Venus Emissivity, Radio Science, InSAR, Topography, and Spectroscopy*, or *VERITAS* (Latin for "truth") mission is an orbiter. It will use its box of tricks to study Venus's surface and map it out in unprecedented detail. It would peer at Venus's innards. Its mysterious geologic workings, and its volcanism, would be better known than ever before. We would learn how the climate, surface, and viscera of Venus are connected.

Smrekar, the planetary geophysicist from NASA's Jet Propulsion Laboratory, is the principal investigator of the proposed mission. She did her postdoc during *Magellan*, when there were more people in the

Venusian science community. "But most people abandoned it," she says. "The reason I managed to stay engaged and excited about Venus despite the lack of data, is because there's so much feedback between studying the Earth and studying Venus." Understanding one, she says, helps us understand the other. Venus will help us answer one of those top-tier questions: "Earth is habitable. Venus isn't," she says. "Why?"

The other mission, the *Deep Atmosphere Venus Investigation of Noble gases, Chemistry and Imaging Plus*, or *DAVINCI+*, shares its name with that famous Italian polymath. It will deploy an orbiter and drop a probe into the atmosphere, which will record its descent. It will hope to study the surface to an extent, but the meat of the mission involves investigating the sky.

Neither mission will be able to detect life. But one of them arguably has a new, unexpected advantage. "DAVINCI is the one that's going to go into the atmosphere and sample it," says Izenberg, a member of the mission's science team. Its instruments also don't need to be tweaked to allow them to detect phosphine, he says; they can already register it.

Other possible future missions by different space agencies, from Europe's *EnVision* orbiter to a host of Russian robotic explorers, have also gotten a boost thanks to phosphine. But their details are still being hashed out. Right now, 2020 is coming to a close and June of 2021 is circled in my calendar. That is when the next Discovery-class mission, or pair of missions, will be chosen, and there is a decent chance that one of the two Venusian possibilities may emerge victorious. Even if just one is chosen, *DAVINCI+* or *VERITAS* hopefully won't be the only mission to go to Venus for a generation, but the first of many in a long overdue dedicated mission program.

I hope we are heading back to that shrouded world. If we are, life will certainly be on the cards. Scientists have spent so long dreaming of Earth-like worlds, gliding silently through the cosmic ocean, waiting for us to find them. But perhaps our nearest aliens can be found in the skies of Earth's evil—or, perhaps, just very misunderstood and unfortunate—twin. Venus is not some distant planet you will only ever see in fuzzy telescopic images or as artistic representations. Sometimes, when it's late, and my partner and I take Lola for a walk up to the Royal Observatory in Greenwich, a bright light in the sky can be seen sneaking up on the just-set Sun. I've always loved the ability to see another planet so easily. Now I look at that tremendous beacon with a new sense of awe. It's another world, almost certainly one with its own fires raging against the endless darkness, and perhaps one with little apocalypse-surviving critters swimming about in the clouds above.

Whether it's the vivaciousness of its volcanism or the mystery of its microbes, the answers to this planet's puzzles, in some respects, don't matter. What does, says Byrne, is that we try to find these answers in the first place. Whatever we discover will be paradigm-shifting, and will only help us to answer the greatest question of them all: are there other Earths out there? Then, for the briefest of moments, his usual jovial nature and indecorous verbiage gives way to downright seriousness. He almost slams his fist on his table. "And damn if that's not an important question to ask."

ON JUNE 2, 2021, NASA revealed the future of its space exploration program, and in doing so stunned the planetary science community: *VERITAS* was heading to Venus—and so was *DAVINCI+*. Then, just days later, Europe announced that its *EnVision* mission was also joining the party. The ocean world that lost its oceans was getting its very own fleet of scientific sleuths. And, finally, we would be getting some answers.

VIII

THE GIANT'S FORGE

In the summer of 1965, scientists at NASA's Jet Propulsion Laboratory in Pasadena, California, had their work cut out for them. They were already wracking their brains, day and night, over Project Gemini, the agency's human spaceflight program that was a crucial stepping-stone to putting Americans on the Moon. But in the middle of all those scribblings, sketches, plans, schemes, and late-night ruminations, it occurred to them that in the next two decades, something extraordinary was going to happen in the outer solar system: Jupiter, Saturn, Uranus, and Neptune were about to be orbiting in such a way that a spacecraft, if launched at a precise moment, could visit all of them on its way into interstellar space.

All four planets had never been seen before other than through telescopes on Earth. If a box of scientific instruments could take advantage of the gravity of these worlds and use it to slingshot past every member of the quartet, humanity would get its first close-up views of the gas giants, Jupiter and Saturn, and the ice giants, Uranus and Neptune, colossal worlds of storms, rings, and alien moons. It would be a prize unparalleled in scientific history. This planetary alignment—not so much a line, but more of a spiral centered around the Sun—only occurs once every 176 years.[1] If they missed their chance in the 1970s, they would have to wait until the middle of the twenty-second century before they could try a similar maneuver again.

They got to work. Building on the experience of the Mariner spaceships that zipped by Venus, Mars, and Mercury, they conjured up a plan to send not one, but two probes into the dark reaches of our solar system. In the 1970s, with the project in full swing and *Star Wars* about to enter pop culture history forever, they decided they needed to give the mission a catchier name than The Mariner Jupiter/Saturn 1977 Project. It deserved to stand on its own, not sit in the shadow of a former program. So they called it Voyager—and, in 1977, with two separate launches at NASA's Kennedy Space Center, two spacecraft were thrown off the planet.

Voyager 2 left first, in August. *Voyager 1*, although launched in September, would be the first to reach Jupiter and Saturn, hence the seemingly strange numbering. Both would go on to do extraordinary things. *Voyager 2* gave us our first and only close encounter with the ice giants. Much to the delight of everyone, it even flew through a bubble of electrified gas 10 times Earth's circumference that was expelled from Uranus in January 1986, something scientists only realized more than three decades afterward while going through its data spool.[2]

Voyager 1, before becoming the first human-made object to enter the space between the stars on August 25, 2012, gave us a truly spectacular peek at the gas giants. In February 1990, when it was 4 billion miles from the Sun, it turned around to take one last look at home. That photograph, in which Earth is depicted by a single blue pixel against a black canvas, was taken at the request of none other than Carl Sagan. "Look again at that dot. That's here. That's home. That's us," he later said of the planetary portrait. "On it, everyone you love, everyone you know, everyone you ever heard of, every human being who ever was, lived out their lives." The planet, he thought, was in that moment a "mote of dust suspended in a sunbeam."

That may have been the most famous of his soliloquies, but it was far from the only thing Sagan had to say about *Voyager 1*'s grand tour. After the spacecraft barreled through the Jovian system, it took a detailed look at the vaporous titan's four large inner moons.

"Battered Callisto: its craters filled with gleaming ice. Ganymede: bearing the marks of ancient geological activity. Europa: its cracked icy surface, perhaps covering an underground ocean of liquid water," he said, before dropping a bombshell for those not in the know. "And most astounding of all, Io: on Io, we discovered multicolored frozen lakes of liquid sulfur, active volcanoes, and geysers reaching directly into space."

IT WAS MARCH 1979, and Linda Morabito hadn't had much sleep. To be fair, neither had any of the other engineers at NASA's Jet Propulsion Laboratory. The Voyager mission was in full swing. The probes were flying through space at breakneck speeds. Coffee was extremely popular on campus. Doodles on blackboards readily gave way to gibberish. Computers bleeped and blooped, and eyelids remained heavy at all times.

Every now and then, slight adjustments to both spacecraft had to be made to ensure they were on the right flight trajectory, and to guarantee that everything would go according to the $865-million plan.[3] No one wanted to see a couple of pioneering spacecraft slam into a planet or moon by mistake.

After *Voyager 1* had exited Jupiter's kingdom, engineers took a look back at its flight path and its recent observations, hoping to make those all-important tweaks. After 546 days flying through space, and after traveling more than a billion miles from Earth, *Voyager 1* sent back 15,000 high-resolution photographs of Jupiter and its moons. These were for scientists to peruse. Another 93 shots were also taken to help engineers in their navigation of Jupiter.

Morabito, an engineer on the mission's imaging team, was never not busy flicking through photographs. "I came to work to perform the optical navigation image extraction at all hours of the day and night throughout February of 1979, instead of any sleep, and often before the Sun would rise," she later recalled.[4] But the work was never anything less than thrilling because, thanks to the photography of a small spacecraft, she could travel among the stars. On some nights, she would walk through the laboratory alone and reflect on her contribution to the mission. "As I looked at the beauty around me, in stars fading from view, at that still, breathtaking moment in time at JPL, I knew my life's path had intercepted my dream," she wrote.

The moment *Voyager 1* met Jupiter, its scientists and engineers were dumbstruck by the wonders they were seeing, thrilled that they were the first people in history to witness the tumbling maelstrom of the gas giant's clouds and the ballet of its Galilean moons—Io, Europa, Callisto, and Ganymede. These four natural satellites were first spotted by the eponymous astronomer in 1610;[5] upon their discovery, Galileo quickly understood that the sight of moons spinning around their own planet meant that Earth, contrary to traditional cosmological doctrine, was not the center of the universe. That cheeky chap Copernicus really was on to something.

Io, a moon so close to Jupiter that scientists suspected it would have an ancient surface covered in craters, was found to be a yellowish world, something that looked like a cross between a cheese-smothered pizza and an orange that had briefly fallen into a furnace. Craters were nowhere to be seen, suggesting its surface was remarkably young. Something was washing those craters away, but what? This was a new mystery for scientists to chew on, another to add to the vertiginous pile of questions that researchers now had after that historic flyby.

The amount of data coming in at the time, Morabito tells me, "was like a firehose." Workstations became domiciles. "We just lived there, and went home occasionally, and did this work, and every now and then we got a glimpse of the incredible things being seen." Scientists and engineers would, when they remembered to eat, go to grab some food in the cafeteria. "On the monitors, you are seeing something than no human being had expected to see." Jupiter, the giant of the solar system, had been revealed.

On March 9, 1979, four days after the Jovian encounter, the initial excitement gave way to a profound sense of satisfaction. By that point, as *Voyager 1* was heading for Saturn, and most scientists suspected that they had seen the best of Jupiter, Morabito considered not even coming into work at all, hoping instead to catch up on some much-needed sleep. But despite more than a month of self-enforced insomnia, she took the plunge once more. She walked into the lab and sat down at her desk, where some post-encounter images of the Jovian system were waiting to be looked through.

Voyager 1 had managed to capture Io at a seemingly haphazard angle: its cresting horizon bathed in only a sliver of sunlight, which meant that most of the moon was in shadow; the side facing away from the Sun was only dimly lit by starlight reflecting off Jupiter. There wasn't much to see at all, and Morabito and one of her colleagues, Steve Synnott, agreed that the images were essentially worthless.

They could have discarded these images. But, ever the completionist, Morabito performed some image processing on them anyway, just on the off chance that something might emerge from the grainy blackness. She was focused on the positions of the stars behind Io and was doing her best to illuminate and clarify them a bit better. And that's when she saw something: a giant umbrella-shaped collection of light seemingly emerging from behind the moon. At first, it looked like another moon was playing hide-and-seek with *Voyager 1*. She showed the entity to Synnott. When he saw that the object was stretching out 150 miles above Io's surface, he said: "Jesus, what's that?"

Neither of them had any answers. Perhaps, Morabito joked, it could be a flare of energy from Io, a little like the immense outbursts of light and magnetism that occasionally shoot out of the Sun. Whatever it was, it was so large that, despite its appearance, it couldn't possibly have come from Io itself. And, as Synnott left the room to check and see if there were any other images of Io taken at that same moment in time, Morabito was left alone, staring at this strange image. She knew she had found something remarkable, but what was it?

Bristling with impatience, she went to find Synnott to see if he had found or heard anything useful. To her surprise, he declared that, after some thought, the image probably wasn't all that exciting after all. Perhaps it was just a glitch, an errant artifact in the photograph. Dejected but determined, Morabito showed her image to astronomer and camera expert Peter Kupferman. "Oh my God!" he said, as he pressed his nose up against the screen. He asked Morabito if she could call Andy Collins, an envoy for the scientists and engineers working on the mission. More and more experts soon arrived to have a gander, and excitement began to echo around the building. Collins had suggested three possibilities: that the object was a newly discovered moon of Jupiter; that it was a moon of Io, a moon's moon—a submoon, or a moonmoon,[6] depending on which scientist you speak with; or that it was, well, something else.

Channeling Obi-Wan Kenobi, they agreed on one of the options: that's no moon, because something that big would have been clocked from Earth. Maybe, said Collins, it's a cloud emerging from Io. But how could this be?

Coincidentally, Synnott had just had lunch with some scientists who were discussing a curious new paper. The study in question, published just as *Voyager 1* was flying by Jupiter, had boldly predicted that

Io may not be an old, cold, dead, crater-filled moon at all—instead, it could be covered in volcanoes.

Erupting volcanoes.

AS A FAIRLY CLUMSY PERSON, gravity is my enemy.

Earth has a decent amount of mass, so it has a decent gravitational pull. It is far too potent a pull for a physically inelegant individual such as me to handle. Gravity isn't just inconvenient—it's also extremely weird. The same force that allows defecating birds to deface parked cars also punches holes through the universe. When a star a few times more massive than our run-of-the-mill Sun dies, it explodes catastrophically and leaves behind a corpse so dense that it turns into a black hole, a light-eating whirlpool that consumes and obliterates everything that gets too close to it.

Gravity also alters the flow of time. This applies to any object with mass, including the Earth. People living at sea level are closer to Earth's center of mass, so they experience a more intense gravitational pull than anyone living at the top of a mountain. Although no one would notice, the passage of time atop the mountain is faster than at the coast, an apparent wrinkle in spacetime that you can actually measure with ultraprecise clocks. That means mountain-dwelling humans age a teeny tiny fraction of a second faster each year than their sea-level cousins.

This effect becomes truly noticeable, to a horrific degree in fact, if you approach a black hole. In the movie *Interstellar*, several astronauts aboard a spacecraft head down to an ocean world orbiting a black hole named Gargantua. One astronaut remains aboard the spacecraft. The explorers are on the ocean world for less than a day. But as they are so close to that shadowy monster, every hour spent splashing about translates to *seven years* of time passed back on their spacecraft. When they return to the ship, the crew member who remained aboard reveals that 23 years had passed. Parents back on Earth had died. Children had become adults. This isn't science fiction, but fact: gravity really would have this terrifying effect in real life.

Between the extremes of time-dilating black holes and me spilling coffee everywhere, there is a middle ground. The Moon may not be that massive compared to Earth, but it's big enough and close enough that, as it spins around us, its gravity tugs at the oceans, creating tides. The Sun, being enormous, helps make tides on Earth too, but the Moon is the main player here.

You might think that you need water for tides, but Io, whose water sped into the dark long ago, seems unlikely to have any. On March 2, 1979, a study[7] led by Stanton Peale of the University of California, Santa Barbara, was published and explained that this aquatic absence didn't matter. Io has tides, because Io isn't alone. Io's orbit is very close to those of Europa and Ganymede. And Peale and company noted that these three moons had a very specific foxtrot: for every single orbit of Jupiter made by Ganymede, Europa made two, while Io made four. This mathematical rat-a-tat is something known as an orbital resonance, a type of gravitationally perturbing set of celestial circumnavigations.

Everything with mass warps reality. It's a little like putting a golf ball and bowling ball on different parts of the stretched-out skin of a balloon. The greater the object's mass, the deeper the depression it makes on the balloon's skin. Planets and stars do the same to the fabric of spacetime, with more massive entities making deeper depressions known as gravity wells.

The Sun contains more than 99 percent of the mass of the entire solar system; consequently, it makes a serious indent in spacetime, creating an enormous gravity well. The planets orbiting it are really just falling into the well. But they are moving at such incredible speeds while they do so that they keep from tumbling into the Sun's deep pit by zooming around the well's inner walls, like Olympic cyclists going around the curved walls of a velodrome. In other words, Earth is orbiting oblivion, but it is spared from destruction thanks to the merciful laws of physics.

The gravity wells of these planets, and their moons, overlap with each other too.[8] Jupiter, a true giant, has a huge gravity well that tinkers with the orbits of the asteroid belt hovering between itself and Mars. Its immense gravitational tendrils push and pull at those rocky and metallic shards of destroyed would-be worlds, creating lanes on the asteroid highway around the Sun. Some lanes are packed with asteroids, while others lack them. Neptune and Pluto are also sharing wells: the ice giant makes three orbits of the Sun for every two of Pluto's, a waltz that ensures both orbits remain stable.

You can get bad resonances, too. Long ago, many icy moons were orbiting around Saturn, with their overlapping gravity wells causing a chaotic game of tug-of-war. Instead of creating stable orbits where everyone shared the space around the gas giant fairly, the moons began to undulate wildly. Some smashed into each other, and their gelid debris fell too close to Saturn's gravitational whirlwind to allow them to bunch back together under their own gravitational fields and make a new set of moons. Instead, some scientists suspect, they became Saturn's rings.[9]

The samba between Io, Europa, and Ganymede could have ended just as destructively too, if it weren't for their mathematically sublime orbital configuration. That 4:2:1 groove[10] ensures that their orbits are stable, but that doesn't mean they all emerge from the dance unscathed. Europa and Ganymede pull at Io. Despite their smallish size, their combined effort causes Io to wobble back and forth, making its orbit more elliptical. That means it gets closer to Jupiter at some points, and experiences a greater gravitational pull, and farther away at others, where the gravitational pull is weaker. The constant to-ing and fro-ing makes the rocky surface of Io regularly rise up and down by 330 feet, the same height as a 24-story building.

If that sounds strange to you, then congratulations: you're a perfectly normal human. This sort of trampolining happens on Earth, but it's the oceans that are going up and down, making high and low tides. On Io, the tides are happening in *solid rock*. This infinitely repeating mayhem repeatedly squashes Io's innards together, creating a vast amount of frictional heat.

"What happens if you have a basically limitless energy? That's sort of what's going on inside Io, it's limitless," says Jani Radebaugh, a planetary scientist at Brigham Young University who, among other things, loves to study Io. Back in 1979, Peale and company wondered the same. They did the math and realized all that frictional heat is probably going to make a lot of magma. Ultimately, they predicted that "Io might currently be the most intensely heated terrestrial-type body in the solar system." It would not be a dead orb, but a world covered from top to tail in volcanoes—ones that are erupting like there was no tomorrow.

And just one week after their paper was published, a handful of people at NASA's Jet Propulsion Laboratory were staring at a strange new image of something flying off the surface of Io. Morabito remembers Synnott, after hearing about Peale's paper, coming back from lunch to her area with a serious attitude. That image wasn't unimportant. It was about to change the world.

There was no way to know for sure that they were looking at a volcanic cloud unless a matching volcanic landform could be seen on the surface. And so, the hunt began. The image was processed further over the next few days, and it looked like the plume was coming from the side of Io facing them, not from beyond the horizon, which helped narrow things down a little. Other detectives started looking through additional images for any signs of something suspect.

Worries about finding the volcano responsible quickly faded on the morning of March 12, while Morabito and Synnott were preparing for a meeting. Suddenly, the phone rang.

"You'd better get up here," shouted Kupferman. "They've found volcanoes all over the place!" Morabito could hear people screaming with excitement in the background. Kupferman cried, overcome with joy.

It turns out that plumes shooting up from volcanoes all over the moon had already been captured by *Voyager 1*, but until now no one had taken the time to study and identify them. Morabito had noticed and recognized the very first: an umbrella cloud of frozen volcanic matter, shooting up from a volcanic cauldron far below. There was even a second plume in that groundbreaking image of hers, one reaching up across the boundary of light and dark on Io, projecting itself above the shrouded surface. And now everyone could see them, on all kinds of other photographs of Io, plain as day.

Turns out that the Peale study was the most serendipitously timed paper of all time. Their forecast, one made by looking at little more than the ballet of moons, was right on the money. And on June 1, 1979, the discovery was formally announced to the world in an issue of *Science*,[11] with Morabito, Synnott, Kupferman, and Collins's names attached. It declared that alien fires, silhouetted against the night, had finally been caught in the act.

Earth's volcanoes were not special after all. Active volcanoes had been found on another world, captured in the middle of several spectacular, sky-piercing eruptions. Morabito's volcano, the one responsible for the very first off-world plume seen by human eyes, was given the name Pele: the Hawaiian goddess of volcanoes. The second plume hiding on the horizon of Io in that very same image was attributed to the volcano Loki, a mischievous Norse god of, among many other things, fire. Over the years, other Ionian volcanoes were named in a similar manner, from Prometheus, the Greek Titan who stole fire from the gods to give to humanity, to Surt, the Norse fire giant.

Forty years later, as *Voyager 1* drifts alone 14 billion miles from Earth,[12] Morabito still revels in her remarkable find. She remains humbled by the beauty of what she had seen and sounds privileged to have been the first to have seen it. "To have been any part of that, any part of it, I'm truly, truly fortunate," she says. It will always remain "the stuff of dreams."

THE DISCOVERY OF erupting volcanoes on Io, a world a tad bigger than the Moon, inspired a generation of scientists. There are plenty of deep and meaningful scientific ramifications attached to this revelation, and we will get to those. But the fact that these volcanoes were erupting so vigorously that they were literally firing stuff *into outer space* was in itself a life-affirming moment for many.

Ashley Davies, a volcanologist at NASA's Jet Propulsion Laboratory, is a hardcore Io fan. He speaks about its volcanism with the same level of whoop-inducing, fist-clenching excitement that Lando Calrissian had when he flew the *Millennium Falcon* out of the exploding second Death Star. "I was always interested in astronomy," he says. "As a schoolboy, I had a telescope, [and stood] out in the freezing winter nights on the front doorstop with the telescope set up, frost forming on the tube, looking at Jupiter and Jupiter's moons."

In 1980, the cataclysm at Mount St. Helens cemented his interest in volcanoes. He went on to study both astronomy and geology at university, and it turns out that he had pretty much the same reasons for fixating on volcanoes as I did: most geology takes forever to happen, but volcanoes can explode right in front of you. "Volcanology's great. When it comes to geology, I just don't have the patience to be a sedimentologist," he says, chuckling. "Volcanoes are right there in your face!"

The worlds of space and volcanoes inexorably came together because, as it turns out, there are volcanoes in space. Voyager was an instrumental part of his upbringing because it brought Io into the picture. He was a schoolkid at the time of the flybys, and "up until that point, all the outer solar system satellites were thought to be these cold, dead ice balls. And then Voyager goes by and on this small body we got these incredible, powerful erupting volcanoes," he says. "And they're all over the place."

The Voyager program may have caught the first glimpses of Io's volcanoes and these towering plumes, but it was NASA's *Galileo* mission,[13] which arrived at Jupiter just before Christmas in 1995 for a nearly eight-year residency, that blew the lid clean off. *Galileo* was equipped with a magnetometer, an instrument that detects magnetic fields. Magnetism and electricity are two parts of the same force—electromagnetism—so this instrument could also feel out electrical currents far away. And as it zipped by Europa, Callisto, and Ganymede, it detected electrical currents flowing below their icy surfaces. A very similar sort of current on Earth is found in its oceans. That meant that these frigid lands were ocean worlds.

That was an undeniably dramatic revelation. But Krishan Khurana, a planetary scientist at the University of California, Los Angeles, was taken further aback when an electrical current was also heard whispering inside Io. Solid rock conducts electricity, with some rocks being more conductive than others. But Io's electrical signal didn't match up to anything solid. "I could see I needed something very conducting," he says. "So I started reading up on magmas, and it was very clear to me that their conductivities are not that different from Earth's ocean. That was my aha moment."

Io is an ocean world too. Except its tides are so immense that it isn't just melting a little bit of rock, but an entire layer of Io. A huge section of its mantle is an ocean of molten rock.[14] It may not be one big pool of liquid minerals. Khurana reckons it's a bit like a slurry, like the molten tar that's used to pave roads, a mixture of solids and liquids. Maybe the melts are moving through holes in a rocky sponge, a bit like a souped-up version of Earth's magma reservoirs. In any event, there is certainly an "ocean" of magma down there, one at least 30 miles thick.

With so much magma to go around, no wonder *Galileo* saw nothing but volcanoes. Flying past Io several times, it captured stunning hi-res images of the moon's freakish surface. And with no life, plate tectonics, or much of an atmosphere to speak of, scientists could see Io's volcanoes completely unimpeded. As Davies wrote in a 2001 paper:[15] "To a volcanologist, Io is a paradise."

Rosaly Lopes, a planetary scientist at NASA's Jet Propulsion Laboratory, tells me how her perusal of Io through *Galileo*'s eyes earned her a place in a rather special book. Finding volcanoes using *Galileo* was pretty easy. "In infrared, the volcanoes just popped up," Lopes tells me. "I started finding all these hot spots, as we call them on Io, the active volcanoes, and my colleagues would joke with me, 'oh, you've found another hot spot.' And then someone—I don't even remember who—they started saying that I should be in the Guinness Book of World Records because I'm finding so many hot spots."

She didn't take that idea seriously until she took a postdoctoral researcher from England under her wing. Upon hearing of the tongue-in-cheek suggestion, he told Lopes that, by sheer coincidence, he went to university with a friend who now works for the Guinness Book of World Records. He contacted said friend, who contacted Lopes. And after the adjudicator ascertained that her claim was legitimate, she was emblazoned in the hallowed tome in 2006, honored for discovering more active volcanoes anywhere in this cosmos—71 of them!—than any other human, a title she still holds to this day.

Thanks in part to Lopes's work, scientists know that almost every volcano on Io is a hole. They look like larger versions of the calderas we see on Earth. They can have ridges that suggest they may form through progressive collapse, a bit like what happened in 2018 at Kīlauea's summit. But no one's quite sure what's going on, so on Io they call them paterae, named after the saucers ancient Romans used to hold their libations. What's more, their cup frequently runneth over, with lava flows near-constantly flooding their floors, spilling out onto the land, and rushing over the moon's horizon. Thanks to all this volcanism, the lay of the land can change fast. "It's a volcanologist's paradise," says Lopes, "and a cartographer's nightmare."

If you fancied bounding around on Io, it probably won't be the lava that kills you. Like the Moon, Io doesn't have the gravity to hold on to enough gases to make an appreciable atmosphere. The surface temperature hovers at around –235 degrees Fahrenheit in the sunlight, dropping to –270 degrees in darkness as Io swings behind Jupiter. If you were to walk on the surface of Io unprotected, in or out of a Jovian eclipse, you would quickly freeze to death as your blood, experiencing a near vacuum, would boil.

Despite the perma-winter, Io's lava keeps on trucking for hundreds of miles without turning into a chilly rock. Although some candidates have been identified, no one's yet seen clear evidence of a lava tube on Io, which, like on Earth, would keep lava nice and toasty, allowing it to flow over large distances. But it's pretty likely that Io is using this cheat to keep pasting lava all over its surface. And it does so constantly, meaning that this 4.6-billion-year-old moon has a surface a mere 2 million years old, on average.

A surface flooded over and over again in lava would be flat and featureless. But on Io, so much new magma keeps gushing up through massive, diabolical esophaguses and onto the surface that the crust gets squashed down. This, says Davies, sometimes snaps the crust upward, forming sudden mountains 12 miles high. On Earth, plate tectonics acts as a sculptor. On Io, you get the geologic equivalent of Hulk smash.

Every lava-filled paterae has something to shout about, but Loki Patera quite literally outshines all others. This bright flare on Io appears to be a lava lake, one that puts Halemaʻumaʻu to shame. Katherine de Kleer, a planetary astronomer at the California Institute of Technology, spends a lot of time using telescopes on Hawaiʻi to stare at Io, and

Loki Patera never fails to impress. It contains a pond of lava the size of Wales. If you stood on its shores, she says, "you will not see anything but lava all the way to the horizon."

Like Earth's lava lakes, says Davies, its surface cools and solidifies; frozen chunks then founder and sink, revealing fresh lava below. It is so huge that waves of lava drifting across the lake in different directions have been seen all the way from Earth.[16] Although its heat output fluctuates depending on how much of the lake has a crusty cap, Loki Patera is always cooking. "On average, it puts out something like 10 percent of Io's total thermal emission," says Davies, grinning and clenching his fists as he speaks. "Which is crazy! This one spot on Io." There is nothing comparable on Earth, which is great news. If Kīlauea suddenly started outputting 10 percent of the planet's heat, part of Earth would just melt off.

Loki Patera's lava lake, probably the top of a mantle plume, isn't merely the most extreme radiator on a world full of volcanoes. It's the biggest single volcanic emitter of heat in the solar system—an inconceivable incinerator.

But even that pales in comparison, at least in my mind, to those plumes first seen by Morabito. Eruption plumes on Earth reach heights of a dozen or so miles. On Io, they reach heights of hundreds of miles. The plume she first saw, emerging from Pele, extended 150 miles above the surface. Another, also from Pele, caught at the turn of the millennium by the *Galileo* and Saturn-bound *Cassini* missions, was nearly 250 miles high. That's 45 Mount Everests stacked on top of one another. If this plume erupted from one of Earth's volcanoes, it would reach the height of the International Space Station.

Another was seen by the Pluto-bound *New Horizons* mission in 2007: a series of images[17] of a plume shooting up from Tvashtar volcano, illuminated by the Sun, was turned into a GIF. I remember seeing this animation for the first time during my undergrad years.

It remains the most remarkable space footage I have ever seen, more thrilling to me than any rocket launch or any flyby of another world. It was footage of the giant's forge, launching volcanic jewels into the inky sky for all to see.

Radebaugh also can't get enough of Io's plumes. "It's so nuts," she says, shaking her head.

Pele's is probably the coolest. On the basis of a patchwork of passing spacecraft imagery, this volcano's plume seems to be an unending fountain persisting for decades on end. If you stood on the surface below and looked up, at any time in the past few decades, you'd always find yourself under a particulate umbrella of yellow and purple-ish sulfur and sulfur dioxide crystals.

Unlike Earth's volcanoes, these towers aren't made of bits of silica-rich magma, cooled into ash. The sulfur stuff is the gas dissolved in Io's lava. When it rises to the surface, the magma decompresses, and the gases bubble out and rush into the sky. The low gravity means that this matter can get launched with greater ease than on Earth, but the real key here is Io's lack of atmosphere. With nothing to slow them down, these sulfurous gases are sucked up into the emptiness of space at speeds in excess of 2,200 miles per hour, where they quickly freeze.

A lot of the sulfur ices hail back down onto Io, the only weather Io will ever experience. The sulfurous ice forms eerie frosts on Io's surface, which can explode into new plumes if magma from below punches through the crust and vaporizes it. The stuff that escapes Io flies far—perhaps, says Radebaugh, as far as the other Galilean moons. Bits of the salts on Europa, for example, may be Ionian seasoning.

The stuff that does neither helps prop up the thinnest of gaseous shells around Io, a wafer-thin layer of sulfur dioxide gas. This world-bubble is so tenuous that it can only exist when Io sees the Sun: the feeble heat imparted by its light turns some of that sulfur dioxide frost into a gas, where it rises and forms a vaporous envelope. But every single time Io drifts behind Jupiter and the moon falls into a wintery darkness, the entire envelope collapses like a deflating balloon.[18]

All of these features cast Io as an extremist world. But its so-called outburst eruptions are beyond extreme. Every so often, there is an explosion on Io, a mammoth energy spike that lasts for a few weeks or so, sometimes doubling the moon's brightness. These aren't plumes or lava flows. "So far, the best theory is that these things are just massive lava fountains," says de Kleer. They can be five times higher than the Statue of Liberty and many miles long, a curtain of fire the length, and height, of a small city. If you placed Kīlauea's fissure 8 in the middle of one of these Ionian incinerators, its own fountaining wouldn't register, a bit like turning on a flashlight while standing in front of a nuclear explosion.

In 2013, three of these outbursts were seen on Io from Earth. "These were just titanic eruptions," says Davies. In less than a week, these eruptions were oozing up to 25 times more lava than Kīlauea's entire 2018 grand finale could manage. "The volumes of lava erupting are like nothing we've seen on Earth—in human history, anyway."

But an outburst in 2001, from Surt, remains the one to rule them all. This eruption's power, which at its peak matched that of all of Io's active volcanoes combined, was clocked at 78 terawatts. This enormous number means that, at its peak, this short-lived but extremely flashy eruption briefly exceeded the current power-generating capacity of the United States by a factor of 65. This number proved so eye-watering that it made it into the 2002 Guinness Book of World Records as the most powerful eruption ever witnessed anywhere.[19]

Io also holds the record for the most voluminous eruption ever seen anywhere. Peaking in mid-June 1997, an eruption at a volcano named Pillan Patera chucked out 13 cubic miles of lava in just a few months,[20] dumping a 30-foot-thick layer of liquid flame over an area roughly the size of Delaware. With so many hits, Io's volcanoes are the Rolling Stones of volcanology. And like this revolutionary rabble, I don't know how to describe Io's superlative nature in a way that would do it justice.

Fortunately, Khurana steps up to bat for me. It seems that Io, unlike the Rolling Stones, would not be a delight to experience in person. A trip there would be about as enjoyable as fending off a pack of hippos with a piece of cauliflower while someone pours scorching coffee into your ears. Io is, he says, "something like Hades. Hell. I've been to some volcanic provinces, and you are maybe 10, 15 miles away from a volcano, and you can start smelling it. And you can be very sure the smell is overpowering everywhere." Presuming you could smell the haze on Io without perishing, he adds, "it's not something you'd ever want to encounter."

If you were close enough to an Ionian furnace, the thermal blast from a nearby lava lake would lightly broil your face, while the lethal chill at your back would draw the heat out of you through a trillion Siberian syringes. If you managed to survive all that, the radiation bombarding the unprotected surface would speedily fry your DNA. Io, says Davies, is "utterly, totally, and completely lethal to human habitation."

But—and this is a big but—if you were sufficiently shielded, and you weren't clumsy enough to fall through a thin bit of crust and into fresh magma, you would see a world dyed with unearthly hues. "Every color would be represented. The contrast would be just amazing," says Khurana. "In terms of beauty, it must be stunning."

AS EVER, there is plenty scientists don't know. Although they quickly agreed that tides were definitely powering Io, it wasn't clear where inside Io the heating was happening. That's like going over to someone's house for dinner, eating the smorgasbord of amazing food they serve, but you never once catch sight of the oven or stove. That would be unsettling—and Ionian scientists were in much the same mind-set.

They reasoned there were two options. If the tidal heating was happening higher up in the mantle, perhaps where a magma ocean sloshes about, then you should see more volcanoes near the equator. If the tides are stirring matter deeper down, Io's volcanoes should prefer the poles.

A few years back, a bunch of scientists, including Lopes and Radebaugh, used a new geologic map of Io to see where all its volcanoes were. The clustering of volcanoes suggested the tidal heating is happening higher up, but there was a problem: the volcanoes weren't in the places the heating models predicted. Many of them were offset to the east by hundreds of miles.[21] Why? There are some ideas, but the fact is that no one really knows what the inside of Io looks like. NASA's *Juno* mission, whose priority is peering at Jupiter itself, is making sweeps of Io's polar regions, and some of its technical doodads may help rule out some possibilities. But until Io gets its own dedicated mission, questions like this will linger unanswered.

It also turns out that it isn't easy[22] to tell the temperature of the lava on Io. You need to catch it just as it erupts or else it will quickly cool hundreds of degrees. And without a mission continually in orbit above Io, scientists can only get snapshots of the lava's eruptive temperature at extremely low resolutions, says Radebaugh, forcing them to use models to give best estimates.

But new lava often seen on Io is humming at temperatures matching that of freshly erupted Hawaiian basalt. That means that, when it first erupted on Io, it was likely hundreds of degrees hotter—and some powerful eruptions caught in the act seem to corroborate that notion. It's possible that the lava on Io erupts at a jaw-dropping 3,000 degrees Fahrenheit. If you fell into this, the water in your body would vaporize so quickly your skin would practically explode.

This is easily the hottest lava known to science. Earth once had this satanic lava back in its formative days, when the planet's innards were hotter than they are today. In this respect, Io is letting us peer back through time. It is showing us a possible configuration of our homeworld when it was still trying to piece itself together, a history obliterated by plate tectonics. This, for volcanologists, is the deeper value of Io. They may not know that much about this moon with any certainty, but it is abundantly clear that it has so much to teach us about the way worlds work, including our own.

"With planets, you're to some extent always trying to use things that are happening on the surface to infer something that's going on in the interior or something about the history for the planet, and it's always complicated," says de Kleer. There is always something occluding the goods, from the atmosphere to water to biology.

Io, though, is volcanism in its purest form. Even from Earth, hundreds of millions of miles away, scientists can see the ongoing construction of the skin of a rocky world without any geologic impediments, the creation of the canvas onto which everything else is painted. "We're seeing processes that help form the surfaces of all the terrestrial planets and the Moon, hundreds of millions if not billions of years ago," says Davies. "And this is the sort of thing that's happening on Io right now. So it's this great laboratory, right? For going back and understanding Earth's distant past, the Moon's distant past, Mars's distant past."

For many Io fans, the excitement of studying it, and infecting others with their own delirium, is reward enough. "It almost always is still just a crazy, wonderful place to look at," says de Kleer. Sometimes, she tells me, Io has a dull day—it isn't as incandescent as it could be. But to her students looking upon Io for the first time, even the most humdrum of snapshots can blow their minds.

From student to scientist, simply watching this preposterous moon cavort around Jupiter is an education. It underscores a vital truth: that Earth may be normal to us, but the universe has other ideas. With phenomenal flamboyance, Io shows us, says de Kleer, just how different things can be. You don't need radioactivity or imprisoned primordial heat to keep a world alive. All you need are tides. And, perhaps most important, Jupiter's strangest moon prepared us for something even stranger: volcanic eruptions whose night fires hold no fire at all.

YOU'D BE HARD-PRESSED to find a spacecraft more beloved than *Cassini*. Designed to unlock the secrets of Saturn, NASA's robotic explorer spent 13 years swooping around the solar system's most beautiful planet. It orbited Saturn 294 times, made 162 flybys of Saturn's moons, and it even threw a European probe at Titan, a mysterious moon that not only has a thick atmosphere but also has lakes, seas, rivers, and rain made of liquid methane and ethane. And when it exhausted its fuel supply, in order to not crash on any of Saturn's moons and risk any sort of biological contamination, it was ordered to plunge to its death in the gas giant's atmosphere—but not before it made 22 death-defying dives between Saturn and its icy rings, studying them as it did so.

Cassini burned up in the skies of Saturn on September 15, 2017. Scientists and science journalists mourned its loss.[23] But the melancholia over its inevitable death was eclipsed by the achievements of its life: around 4,000 scientific papers were written about its cornucopia of discoveries. And according to Linda Spilker, *Cassini*'s project scientist and a Voyager veteran, one of its most astounding finds came about when it paid a visit to a moon named Enceladus.

From Voyager's perspective, Enceladus, a poky place no larger than the United Kingdom, was a mirror made of ice. Other icy worlds were fairly reflective, but Enceladus looked like it was shedding starlight. When *Cassini* swooped in, it found an icy shell concealing yet another huge saltwater ocean. It also found that the region around its south pole was essentially free of impact craters—an extremely young part of an already youthful surface. Vast chasms nicknamed "tiger stripes" were present. Icy material was gushing out of them and into space at 800 miles per hour. "We were all, wow," says Spilker. "A moon that shouldn't be active by everything we think we know. It was so tiny that the thought was that it should have been frozen solid." And yet there it was, flinging water-ice into space and providing the building blocks for Saturn's bright, icy E Ring.

The water was being sucked out of the chasms by the vacuum of space, while bubbles trapped in the icy slush below were likely expanding violently whenever the slush found a pathway to the surface. "It's kinda like shaking up a champagne bottle then taking your thumb off the top," says Spilker.

That part wasn't too complicated. What was harder to explain was where the geysers were coming from. If this moon is several billion years old, then how has it got enough heat still trapped inside it to keep an entire ocean liquid? The salt content of Enceladus's ocean certainly helps stop it from freezing over, but was it sufficient to keep it liquid for eons? There's also no way that it still has enough heat left over from its formation or from the decay of radioactive elements to keep things snug. What's really going on here?

It soon became clear that there wasn't going to be a simple solution. When the intrepid mechanical adventurer flew through one of those jets in 2008, it found an array of organic materials and nanoscopic grains of silica dust. "The interesting thing about those grains is that they can only grow and form in water that's [194 degrees Fahrenheit] or above. So you're almost at the boiling point of water," says Spilker. *Cassini* also smelled an excess of hydrogen gas. And both are found coming from a very specific environment on Earth: deep-sea hydrothermal vents. So not only is there a presumably ancient liquid ocean down there, but there is also volcanic activity on the seafloor. What sorcery is this?

Cassini's beady eyes found a clue to the conundrum around the moon's south pole: the tiger stripes weren't static and cold. They were relatively warm, and the jets were firing out of the warmest spots.[24] Not only that, but the entire region appears to be in constant motion, with the tiger stripes opening, closing, stretching, and warping. With plate tectonics out of the question, scientists suspected gravitational tendrils must be trying to rip it apart, and it just so happens that Enceladus is in orbital resonance with the moon Dione.

The Saturnian system is complicated, with many moons all conjoined in a gravitationally complex, messier dance than the boogie between Jupiter's Io, Europa, and Ganymede. Though the magnitude of its role is uncertain, tidal heating is most definitely playing a part in keeping Enceladus warm. The very same invisible force that makes the hottest lava on the most volcanic object in the solar system may also be responsible for making hydrothermal vents and maintaining a liquid ocean on glacial Enceladus. And if that heat is melting solid icy matter, opening cracks and allowing said molten ice to erupt into space . . . well, what exactly would you call that?

Some call them geysers. Others call that cryovolcanism: volcanic activity, but with ice and water, not rocks and magma. The coldest lava, perhaps, in the universe, chilly enough to not melt anything, but freeze it on the spot.

THERE'S A BIG ROCK in the asteroid belt named Ceres.[25] It contains a quarter of the entire mass of all those asteroids. It is not anywhere near the scale of any of the solar system's planets, and it doesn't have an orbital path all to itself, so it's not a planet. But it's pretty big, so astronomers called it a dwarf planet. It looked quirky, so NASA pointed a spacecraft named *Dawn* at it in 2007 with instructions to go and have a look. When it arrived in 2015, it found a grubby ice ball covered in dark plains, weirdly bright spots, and mountains made of ice.

One of those mountains, Ahuna Mons, was 2.5 miles high and shaped like a dome. Its freaky smoothness and shape suggested it was pretty young. And it was part of a wider family of other domes[26,27] scattered all over Ceres's mostly gray surface. It was hard to tell from orbit, but from what *Dawn* and its scientists could fathom, there was only one explanation for these domes that made sense: these were ice volcanoes.

A paper[28] from 1971 may be the first to suggest that ice volcanoes could exist. It doesn't use that term, but it does suggest that some of the natural satellites of Jupiter, Saturn, Uranus, or Neptune, if fractured, "would permit violent eruptive release of water and ammonia onto the surface." That is pretty much what ice volcanoes, or cryovolcanoes, are thought to be: icy hills or mountains that were built by the eruption of molten, viscous ices made up of water, methane, ammonia, and other liquids that can freeze up into a solid—cryolavas, in other words. And those lavas are pretty bizarre. Methane is the same stuff that burns on a natural-gas stove. Ammonia will be most familiar to those who own cats: it is responsible for the nose-violating, acrid stench that seeps from old feline urine, whose urea component is being broken down by bacteria.

But no one knows for sure what an ice volcano looks like or how it erupts, because unlike the explosive geyser-like ice eruptions emerging from the tiger stripes of Enceladus, we haven't caught a suspected cryovolcanic construct in the act of squeezing out icy lava. That means we can't be certain we've ever seen a bona fide cryovolcano.

Some scientists, though, are convinced we've found a fair number of them. A few years ago, Lopes suggested that some flow-like, reflective bits on Saturn's moon Titan were made by cryovolcanoes. She wrote a paper about the sites, and right afterward, two of her colleagues wrote their own paper arguing that there was no ice volcanism on Titan, and that anything suspicious could be explained away by other, less exotic processes. And then, Lopes tells me, *Cassini* found a big, cold mountain on Titan. Scientists, inverting the qualities of the ring-destroying furnace from *The Lord of the Rings*, named it Doom Mons.

Lopes was fairly convinced that it was an ice volcano. But the same colleagues disagreed, and for several years they had a friendly back-

and-forth, with everyone standing firm on their side of the fence. One of them, Jeffrey Moore, was a member of the *New Horizons* spacecraft team. The same year that *Dawn* arrived at Ceres, *New Horizons* reached its own dwarf planet: Pluto, a true ice ball at the boundary of the solar system. "And then," says Lopes, "they found Wright Mons," a young, 90-mile-wide mountain of ice with a hole at its summit.

Shortly after, there was a conference about the Pluto flyby. According to Lopes, Moore started his presentation by saying: " 'Rosaly Lopes must be laughing,' because his talk was proposing that Wright Mons was a cryovolcano." It was soon joined by Piccard Mons, another possible Plutonian ice volcano.

Again, no eruptions have ever been seen from any of these icy palaces. Maybe they are just funky-looking lumps with very curiously positioned impact craters at their peaks. Sure, they walk like a duck, but do they talk like a duck? Adeene Denton, a planetary scientist at Purdue University, tells me that most scientists look at these mountains and say: "It has to be volcanism, right? How else are you going to build something that big?" Others may say, how can you be sure? And they'll respond: "Well, unless you have a better idea, this is a volcano."

The big problem, other than the lack of actual ice lava eruptions, is that the laws of physics seem to make the existence of ice volcanoes improbable.

Imagine you have a glass of water, and you plonk some ice cubes in it. The solid forms of most materials sink into their liquid form because the solids are denser than the liquids. But ice doesn't play by the rules. Its molecules are not squashed together, but instead spaced rather far apart. That makes ice less dense than water, which is why ice cubes float in your drinks.

Hot, liquid magma is less dense than the solid magma around it. That's why it rises and why it can erupt. But "at this point, cryovolcanism is a tough sell for many people because it should be almost impossible to make water go up through ice," says Denton. "It simply does not want to do it." If Wright Mons is an ice volcano, then molten ices needed to travel 60 or so miles through Pluto's thick icy shell to erupt onto the surface. "That's a non-negligible distance for water to travel, right?"

Lopes points to a series of laboratory experiments in the 1990s that mixed water-ice with ammonia. At Pluto-like deep-freeze temperatures, scientists showed you could make a gloopy flowing substance—a small-scale icy lava. "I remember in a TV interview, I was trying to describe this cryovolcanism, and the presenter said: 'Oh, you mean a bit like a Slurpee!' " Maybe that's what an ice volcano is: a mountain retching Slurpee streams stinking of old cat pee.

I'm not exactly sure what would happen if you fell into ice lava, but the smell would be the least of your problems. Upon diving in, the water in your skin, muscles, and organs would freeze and expand, splintering your every cell and tearing you apart from the inside-out while also preserving you as a macabre statue prone to cracking. The heat of your body may also violently vaporize some of the cryolava into its constituent gases. Neither you nor the segment of lava you tumbled into would survive the encounter.

Trippy though they may sound, cryolavas are probably real, but no one has shown how they could erupt and make a mountain. "I don't think it's impossible," says Denton. Despite her doubts, she hopes that peaks forged by volcanic ices exist. Her reasons are, she says, "very selfish": the concept of cryovolcanism is "kickass." Who *wouldn't* want ice volcanoes to be real?

Cryovolcanism—from the ice geysers seen on Enceladus, as well as on Jupiter's Europa and Neptune's Triton, to the very volcano-esque ice mountains of Ceres, Titan, and Pluto—is "definitely a thing," says Lopes. But because scientists have so little data from the cold, dark, and lonely fringes of the solar system, where the Sun is just another speck of light in the night, ice volcanism is largely a closed book. For now, being a full-time cryovolcanologist isn't feasible.

"It's a very difficult problem to approach because we don't have a clear idea of what the physics would be," says Denton. We don't know what keeps icy magmas liquid on all these miniature worlds for millions, if not billions, of years. Ammonia can act as a decent antifreeze,[29] pockets of gas[30] beneath icy crusts could keep liquids below sufficiently warm, and tidal heating, for sure, must help out too in places.

In 2020, scientists examining *Dawn*'s observations of Ceres also came up with another way ice volcanoes could exist: through every planetary scientist's favorite fallback option, a big thing hitting another thing. Andreas Nathues, an expert in planets and comets at the Max Planck Institute for Solar System Research in Germany and member of the *Dawn* science team, walks me through it. Ceres is mostly dark, but it's peppered with extremely reflective patches of what appear to be sodium salts. The best way to explain these patches is through ice volcanism: cryolavas erupt onto the surface, the ices are vaporized by the sunlight into gases that escape into space, and the salty bits of the brines are left behind as the ghosts of eruptions past.

But where are all these cryolavas coming from? A series of papers[31] published in 2020 tried to answer that question by zeroing in on a city-sized crater named Occator, home to the dwarf planet's biggest and brightest patches. This scar was carved out 22 million years ago, and like any powerful impact, it would have produced a lot of heat—in this case, not enough to melt the rocky part of the crust, but more than enough to turn shallow ice into mush. That impact also cleaved off a few miles of Ceres's crust, fracturing the deeper crust and relieving pressure on pockets of molten ice below.

A shallower melt chamber—essentially a cryomagma reservoir—was formed by the energy of the impact itself. Thanks to those lovely new pathways to the surface, cryomagma rose to the surface while a far larger, deeper reservoir of cryomagma brines fed into the shallow chamber. Water may not like rising through ice, says Nathues, but the same process thought to happen on Enceladus may be happening here. "And like a bottle of champagne, if you shake it . . ." Nathues tells me, before displaying jazz hands, mimicking champagne blasting out from an uncorked bottle. All you need is trapped gas and the vacuum of space, and this sort of cryovolcanism works fine.

Most of those cryolavas mostly oozed out like toothpaste, but in some cases, they may have vigorously splashed onto the surface. And when the icy parts sublimated into nothingness, the salty parts remained, forming the bright patches that exist today. Some of the salts appear a little soggy, which suggests that they only very recently erupted. It's not unreasonable, then, to say that these cryovolcanic fissures may still be erupting right now, perhaps just through little episodic trickles. If so, that makes Ceres the shadow version of Io: an active volcanic world, but one whose story is written in ice, not fire.

I know we're talking about science here. But if Shakespeare were alive today, I like to think that cryovolcanism is exactly the sort of thing he was referring to when he wrote that there are more things in heaven and earth than are dreamt of in your philosophy. If ice volcanoes erupting today in the space between Mars and Jupiter isn't magic, then I don't know what is.

I HATE TO ADMIT IT, but these ice volcanoes aren't the most important part of this story. "The cryovolcanism is a signature for what happens at depth," says Nathues. And they seem to signal that all these icy or ice-rich worlds conceal oceans' worth of liquid water.

Denton prides herself in having blown up Pluto countless times[32]—in simulations, of course. She has found that the only way in which Pluto gets its strange hummocky, rippled terrain in one spot and a 1,200-mile-long crater in another spot on the opposite side of the world is if seismic waves from a giant impact were transmitted through a 93-mile-thick subsurface ocean. It's unclear how Pluto would have such a substantial ocean that hasn't frozen up by now, but scientists tend to agree that it does have one. And cryovolcanoes sitting above it only bolsters that possibility.

These oceanic orbs appear to be concentrated in the outer solar system. Ceres positioning on the edge of the inner solar system is odd, but once upon a time it may have been as far out as Pluto. Ammonia ices can exist far from the Sun but would quickly vaporize closer to the stellar bonfire. Ceres has lots of them despite being just behind Mars, suggesting that at some stage it lived on the solar system's distant frontier. And through impacts, radioactivity, and saltiness, it seems to have held onto an ocean of its own for billions of years. There is a debate[33] as to how much water Ceres contains, but this minor slush ball could contain more water than we have on Earth. And once again, those bright spots and its ice volcanoes make more sense if there is a big reservoir of cryomagma down there to sample from a churning ocean.

Closer to home or far from it, water worlds seem to be everywhere. We've probably even missed a few. "We don't know for sure, but there's some thought that maybe some of the other moons of Saturn—maybe Dione, maybe Mimas—maybe they have liquid water oceans, and we just didn't get enough information to know for sure," says Spilker.

On some of these worlds, like Enceladus, there are probably hydrothermal vents. And you remember what loves water and hydrothermal vents on Earth, right? One day, decades from now, we may be sending the descendants of those daring AUVs diving around the Kolumbo seamount on Earth into the oceans of these icy worlds on the hunt for alien life, which could be anything from microbes to octopus-like beings. The way these veiled oceans are kept warm, and the way their ice volcanoes are powered, vary from place to place. But what Morabito found on Io back in 1979 destroyed an old astronomical paradigm.

"At a basic level, it changed our understanding of where life could be supported across the solar system because you're no longer prisoner to the Goldilocks zone around the Sun," says Davies. It showed scientists that you don't need to be Earth—for now, a perfectly placed, hydrated, temperate planet—to create a living, breathing world, one that provides a potential home for life. If you have the right kind of dance, all you need is gravity.

"We're finding a lot of ocean worlds just in our own solar system," says Spilker. "So if you look at all the stars in our galaxy, what are the odds that you have a planet like the Earth in the Goldilocks zone—not too hot, not too cold, where all the conditions are just right—versus having moons around the larger planets we know, like Neptune-size and up . . . there's a lot of these bigger planets and they might have moons where tidal heating may be a factor. You may have more ocean worlds with frozen icy crusts than Goldilocks worlds like the Earth."

If you fired a laser from Earth at a certain part of the sky, the light would take 40 years to travel 230 trillion miles before it reached the TRAPPIST-1 system,[34] a collection of seven rocky planets orbiting an alien star. Some are in the Goldilocks zone, while others are too close or too far to support liquid water at the surface. But, says Davies, these planets are in orbital resonances with one another. Some tidal heating must be happening. It could be negligible. It could also be significant enough to fuel volcanoes—fire, ice, or both—and to sustain liquid water oceans perhaps suitable for life.

To us, Earth's volcanoes, and Earth itself, is our normal. But so far, with our albeit limited search of the great beyond, Earth is looking to be the exception. Primordial heat leaks away. Radioactive decay slows down. But, says Spilker, the peculiar orbits of orbs can provide an endless source of heat. When it comes to keeping worlds alive, perhaps the tides of gravity are the only engines that transcend the tides of time.

FOR MANY SCIENTISTS, the question of where life can be found isn't the be-all-and-end-all of planetary science. "They had one of those books at the library when I was quite little, like six, on Jupiter's moons. And they had a picture of Europa with all its scars on it, what I thought they were when I was a kid," says Denton. "And it said there was a subsurface ocean under here and I was like: this rocks, this absolutely rocks! It's just funny to think about . . . there's so many of them. There's likely so many more. And exploring the intersection between geology that we do understand and things that do behave the same way, and things that don't, and trying to understand why that is, is so interesting to me."

You need functioning worlds in order to support life in the first place, so there's no way we can hope to understand how life arises and proliferates through the galaxy if we don't get a grasp on how planets work in the first place. "I don't care a whole lot about the life angle, necessarily," Denton says, with a smirk. "Maybe there's critters down there, maybe there's not. But I'm honestly more interested in how worlds are alive in a tectonic sense."

Dreaming about these strange volcanic worlds is a joy in less fraught times. But in 2021, with climate change, racial injustice, a pandemic, and economic and political turmoil looming large in everyone's lives, these scientific ruminations can be a refuge. Several of the scientists I interviewed for this chapter were speaking to me from California, which at the time was being savaged by both the coronavirus and wildfires. Most seemed distressed at the state of the world right outside their doorstep. But waxing lyrical about distant volcanoes seemed to provide a temporary haven.

Khurana, though, seemed buoyant, effervescent. I asked him how this was possible.

"For us scientists, we have many lives," he explains. "We live on many of these bodies all of the time. There are months, sometimes years, that I spend on a particular moon or on a particular planet, and first thing when I wake up in the morning is the thought about that planet, not oh, do I need to buy yogurt?" He can work through the ongoing disasters, he says, because he has another life to drift off into.

The scientific quest to understand these volcanic worlds matters. Trying to unpick how the same force can make the hottest and coldest lava in the universe matters. But what seems to make many of these scientists happy brings me a great deal of happiness too: leaving normality behind and taking flight to visit these volcanoes, from the fiery mountains on Earth to those on the worlds beyond. It is why we love reading beautifully written books, listening to moving music, playing exciting video games, or watching thrilling movies. It is why we love stories. It is, for some, why we enjoy writing. We all need a good escape, especially when so much about the world is broken. We all need a good story. And, through science, these volcanoes and planets give us stories that are both true and fantastical beyond our wildest imaginations.

"I think, in general, people who have another world to escape to . . ." says Khurana, pausing for a moment. "I think they're happier people."

EPILOGUE

THE TIME TRAVELER

In the year 79, Mount Vesuvius erupted and engendered the world's most infamous volcanic disaster. The day that rocked the Roman Empire has been spoken about, written about, and televisually recounted so many times, so I'm sure you know the outline of the story: the eruption initially dumped ash and other volcanic debris over the region, including atop Pompeii, causing buildings to collapse; after that, a series of pyroclastic flows and gassier surges blasted across the land, annihilating Pompeii, the port town of Herculaneum, and any other settlement unlucky enough to get in the way. (Around the same time, Emperor Vespasian reportedly said "I think I'm turning into a god" before dying from excessive diarrhea. Clearly, 79 was a terrible year for Roman eruptions.)

Having been so extensively excavated over the centuries—work began on digging up the lithified metropolises in 1748—you would think that any questions about that fateful eruption would have been cleared up by now. But apart from a whole host of archaeological queries yet to be answered, it's not actually clear how thousands of Romans died.

This may sound peculiar. They obviously died from the eruption, right? Of course. The precise way they died, though, remains uncertain. Was it mainly due to collapsing buildings? Flying debris? Ash inhalation? Their organs shutting down due to the extreme heat blast? Being crushed under a panicked, stampeding crowd fleeing for their lives? The fact that such a seemingly straightforward question doesn't have a definitive answer, and may involve different answers depending on where the victim was during the eruption, is both macabre and fascinating. Academics, who hope to find answers to both satiate their morbid curiosity and to better understand the raw, destructive power of violent volcanic eruptions, lean in several directions, but there are two threads that currently stand out.

The first is that many victims were flash-fried to death.[1,2,3] A dark red tint on some of the charred bones of those in Herculaneum looks to be iron-rich. These were perhaps sourced from the victims' blood, sprayed everywhere as their skin, muscle, and fatty tissues were quickly vaporized and their red blood cells ruptured while experiencing extremely high environmental temperatures. Fractured skulls have also been purported to show that the fluids in the victims' brains boiled so vigorously that their heads exploded open. It gets weirder: a glassy substance containing fatty acids and proteins inside the cracked skull of one corpse found in Herculaneum has been suggested to be the vitrified remains of someone's brain, a never-before-seen phenomenon that is only possible if something is heated to an extreme degree and then quenched very suddenly.[4]

This gruesome but instantaneous (and therefore painless) manner of death is seen as unrealistic by other scientists.[5] Experts who study the decay of bones say that you never see exploding skulls inside crematoriums, which get hotter than any pyroclastic flow could manage, and that soft tissues in said crematoriums burn, but they don't vaporize. The glassy brain discovery is gnarly as heck, but it isn't clear what coolant would suddenly chill the victim's melted brain matter into a glass, nor is it understood why no other victims appear to have been cerebrally vitrified.[6]

Another study[7] looked at the bones of 152 people who gathered, and died, sheltering in Herculaneum's stone boat chambers. They found that the bones had plenty of collagen still in them, a protein that breaks down when cooked at high temperatures. Bones baked by high temperatures should also be significantly crystalline, but the victims' bones showed otherwise. Estimating burn temperatures on such old matter is difficult. But previous evidence from the damage to materials in the town have given a range of environmental temperatures of 465 to 1,470 degrees Fahrenheit. Contrary to the flash-fry studies, this paper suggested the lower range was more likely.

Pyroclastic flows and surges unquestionably scorched the region. But in Herculaneum, at least, death was probably not instantaneous. The victims were essentially broiled to death, a horribly painful way to die if they hadn't passed out from inhaling toxic volcanic gases first.

The specificity of their demises is still up for debate. But what remains entirely unambiguous is the soul-shaking nature of the eruption itself.

Pliny the Younger, a Roman lawyer and writer, wrote two letters[8] about the eruption after he had managed to scarper out of there. They are the first detailed descriptions of a volcanic eruption in human history, and some have suggested that they herald the birth of the science of volcanology. The type of eruption that Vesuvius made was later named by scientists after him: Plinian.

The first letter describes his uncle, Pliny the Elder, and his attempts to sail a boat into the umbrella-shaped cloud emerging from Vesuvius so that he could study it. When he received word that a friend was in danger of being swamped by the volcanic avalanches, he changed course, making observations as he went. He arrived at his destination, Stabiae, and tried to calm his friend as buildings violently swayed and rocks hailed down onto them. The waves became unfavorable and, unable to escape, he died on the shore, suffocated by the fire and brimstone that overcame them all.

In the second, Pliny the Younger describes, hauntingly, what he saw across the Bay of Naples as he and his mother fled. He describes earthquakes attempting to tear down the land; the sea being sucked back, stranding sea creatures on the dry sand; a "dreadful black cloud" that was "torn by gushing flames and great tongues of fire like much-magnified lightning"; the same cloud "advancing over the land behind us like a flood."

Resting for a moment, they found that darkness rapidly ensconced them. They could hear people wailing in terror in the blackness, trying to find their loved ones lost in the hellish fog. "Some people were so frightened of dying that they actually prayed for death," he wrote. "Many begged for the help of the gods, but even more imagined that there were no gods left and that the last eternal night had fallen on the world." Pliny said that he put on a brave face at the time but noted: "I was only kept going by the consolation that the whole world was perishing with me."

Eventually, when the darkness cleared, the Sun was struggling to shine through the floating dust and ash, as if an eclipse had occurred. Their surroundings were smothered, right up to the horizon, in a deep, snow-colored veil of ash.

VESUVIUS HAS ERUPTED several times since the great collapse in 79. In December 1631, another eruption buried several villages and killed thousands. The Olympics should have taken place in Rome in 1908, but a Vesuvian paroxysm in April 1906 forced the already behind schedule Italians to divert funding and attention to the damage to the region, giving the games to London instead.[9]

The volcano's most recent eruption, although certainly not its last, was in 1944. By this stage of the Second World War, the fascists were in retreat. Italy had surrendered to the Allied Forces in September of the previous year, who had set up their own forward operating bases in and around the Bay of Naples. In the spring of 1944, Vesuvius woke up.[10] An explosive eruption mimicked the sounds of artillery fire, sending volcanic ash, blocks, and bombs flying. Lava meandered from on high down toward several villages and towns, forcing people to evacuate. Some kids followed the slow-moving lava as it snuck along the road; some roasted chestnuts on it, while others lit cigarettes.[11]

The Allies happened to have set up an airfield for their bombers just a few miles from the foot of the angry mountain. Caught off-guard but unable to look away, many watched the oncoming storm frozen to the spot. A written account,[12] attributed to Sergeant Robert McRae of the 489th Squadron,[13] paints a stark picture. "To look above the mountain tonight, one would think that the world was on fire," he wrote on March 20. "As the clouds pass from across the top of the mountain, the flame and lava can be seen shooting high into the sky to spill over the sides and run in red streams down the slopes." Looking through some binoculars at the summit, he saw "flame, sparks and lava . . . being thrown from the crater like rice at a wedding," a display of pyrotechnics accompanied by bone-shaking rumbling.

Over the next few days, the heavens opened up, and "black stones of all sizes, some as large as a football, fell in great quantity completely covering the ground, breaking branches from the trees, smashing through the tents to break up on their floors, tearing through metal, fabric and Plexiglas of the airplanes." Soldiers made sure to wear steel helmets as they ran for cover, while civilians made do with cooking pans or sizable baskets. Eventually, everyone was forced to evacuate until the worst was over.

On the 25th, after breakfast, McRae and some others were ordered to survey the damage. They found that 88 B-25 Mitchell bombers had been shot through, costing them valuable weaponry and $25 million. Fortunately, they suffered no deaths—"a sprained wrist and a few minor cuts were the only casualties," apparently. And within a week, he proudly wrote, the 340th Bombardment Group was up and running again, so volcanoes and Nazis can jog on.

Although not massively imperiled by the eruption, nerves did fray: the author wrote that several soldiers, as he had, offered a prayer for safety. But, viewed from a distance, the overwhelming theme was one of awe. He simply couldn't believe, nor describe, the eerie, otherworldly, perspective-shifting sight he was witnessing. "This is a sight to be remembered. An ironically beautiful sight. The dark clouds against the red glow with bursts of flame showing between from time to time is quite a picture," he wrote.

McRae's chronicling of the eruption has echoes of Pliny the Younger's account. Despite being separated by millennia, little has changed in how eruptions are experienced on a visceral level. There's no denying that they can kill, and that generates fear. That's entirely understandable. But when seen from a safe or relatively safe distance, as most eruptions are viewed, this fear gives way to astonishment. Before we have even had a chance to think scientifically or rationally, our jaw drops, our eyes widen, and our hearts thunder. We are seeing something unlike anything else, a radiant, seething split in reality.

Time moves on. But volcanoes and eruptions have a timeless effect on our minds, whether we are watching their embers on land, underwater, or in space.

THIS BOOK ISN'T ABOUT volcanoes—not really. It's about time travel. I don't know about you, but the ability to travel through time brings me great comfort. To daydream of other times provides a remarkable sense of control, especially when the world around us has been electrified by chaos.

We only get one go at life. That's a deeply sobering thought. That hasn't made me fear death, per se, but it makes me fret about loss and ruminate on regret. I sometimes stay awake, late into the night, raging at the speedy passage of time. Even during the coronavirus pandemic, where both everything and nothing is happening simultaneously, time is shooting by. I miss my family, my friends, all the little things. I always miss my grandparents, three of whom were taken away so ruthlessly in recent years. On more than one occasion, I've imagined going back in time, just a few years or so, pretending to be my slightly younger self so as to not disrupt the future, and having one more conversation with them, having one more hug, one more chuckle. They would have loved to hear about this book you're reading.

Time can't be rewritten or rewound. But that's okay, because the times we all shared together have become stories. And as the very best of doctors said once upon a time: "We're all stories in the end." Storytelling is a form of time travel. Thanks to those memories, I can relive the days I spent with my friends, my family, my grandparents, whenever I wish. Thanks to the words written down by those who lived before us, from the sergeant to Pliny the Younger, we can visit the memories of others and travel further back in time, through decades, centuries, or millennia.

Our species is still in its infancy. The solar system is our childhood home. But what a fantastic place to find ourselves. It is a home with a boundless library, one full of books whose words are written in volcanic ink. Those tomes are full of stories, of the beginnings and ends of worlds, of the resilience of life, of magical acts of flame and ice. And science lets us read them.

We really do only get one go at our lives on Earth. But these volcanic stories let us walk across the world, and across other worlds, diving in and out of all those memories recorded in volcanic rock, from Kīlauea to Wright Mons, from Ol Doinyo Lengai to Olympus Mons, from Vesuvius to Loki Patera, leaping back and forth, from the distant past to versions of the future, through eons of time—all within a single human lifetime.

How fortunate we are, all of us, to understand our cosmos, and our place in it, through planetary fires. What a gift. What a thrill. What an adventure.

ACKNOWLEDGMENTS

I spent much of the 2010s hopping around the world in the name of academia. Consequently, my friends are everywhere, geographically speaking—often far from wherever I am. I always miss their company, right up to the moment we reunite. But despite the distance, we're always there for each other, and speaking to them during the almighty dumpster fire that was 2020 kept the number of lachrymose moments to a minimum. In particular: Patrick—adventurer, Muse gig companion, and eternal video-gaming buddy; Mitch—geoscientist and conversationalist extraordinaire; Simon—definitely sent back in time to nefariously change the future; Joe—troublemaking, mayhem-chasing adventurer with a golden heart; Jason—an unashamedly childish childhood friend; Kate—artist, curator, and amplifier of all that is silly; Lukas—musician, rogue, and master of madness; Sebi—deep thinker and sonically gifted wonder; Mercan—a fantastic, fiery friend; Judith—the OG Vienna buddy, and dreamer of big things; Daniel—the songful, silent type; Katy—one-third of the evil trifecta, and encourager of daftness supreme; Josh—one-third of the evil trifecta, and proficient producer of chuckles; Rosie—writer-in-arms; Alfredo—smart-ass astrophysicist and the only narcissist I've ever liked; Chris—as heartwarming as he is easy to jump scare; Jonny—a lovely and indescribably ridiculous human being; Kristy—supportive, mad, and marvelous; James—both an acidic wit and a big softie; Tom—wielder of a delightfully dark sense of humor;

Valentina—bonkers, spontaneous, and cool as hell; Jodie—an international weaver of mischief; Mike—fellow enjoyer of geekery and grinning; Christina and Sarah—my very own bad idea bears; Shannon—a one-woman revolution who came out of nowhere; Laura and Tony—dinosaur-prone living legends; Lesley—more family than family friend; George—polymathic, Star Wars–adoring marvel; Franzi—the dinosaur-defeating, science-communicating wonder; Sonny—hilarious erudite and expert winder-upper; Geoff—one of the weirdest, wackiest people I've ever met; Shobhit—*Halo* convert and procurer of endless laughter; Bryce—my extremely brief fake-husband with an affecting soul; Natalie—an author of inspiration; Janine—volcanologist and expert confidence booster; Jess—volcanologist and maverick; Mariana—the best New York tour guide ever; Felicia and Jasmin—fellow infiltrators of the strangest afterparty in the world; Sean—my older cousin, but more like my big brother; and Nathan—my younger cousin, but more like my little brother. You are, all of you, friends new and old, just the best; to the days to come, and all my love to long ago.

My editors, past and present, are the reason I get to rearrange a handful of letters to explain weird rocks and planets for a living. They aren't just brilliant at what they do; they're also genuinely lovely people, the sort you'd be very fortunate to encounter one day. So: Maddie, Victoria, Nsikan, Jay, Kristin, Michael R., Michael M., Lee, Clara, Brian, Rose, Samir, Jeremy, Sarah, Lex, Brian, Robin, Matt, Alex—thanks a billion. I hope this book lives up to your standards!

This book wouldn't exist without my gung-ho literary agent, Lane, nor would all the words sound so delightful without my editors, Quynh and Helen. Thank you so, so much for giving me this life-changing opportunity. Thanks, too, to Justin Estcourt (@jetsyart on Instagram) for making the beautiful art for the book's prologue and epilogue—truly stunning work.

Gratitude must also go to my geography teachers, Mr. Chapman and Mr. Smith, who encouraged my love of volcanoes enough to convince me to study it at university. I also owe much to Shigeru Miyamoto, whose childhood wanderings around Kyoto inspired *The Legend of Zelda* and, inadvertently, my journey into volcanology, as well as some of my happiest childhood memories.

My grandparents, the most supportive, caring, whip-smart people I've ever known, are as missed as they are loved. I never met Grandad Charlie, but he sounded like a right laugh. Nan A. was always thrilled to see me, and she would read every single thing I ever wrote. She would have devoured this book, beaming with pride. Nan M. never stopped singing my praises. She erroneously but charmingly kept telling people I was a professor and hosted the best Boxing Day gatherings in human history. And Nanu (a Maltese slang word for "grandfather," for those wondering) was the cheekiest ancestor anyone could have ever asked for. I carry them with me, wherever I go. My Aunt Jackie, Uncle Yutaka, and Uncle Paul have also shown me nothing but encouragement and support throughout my life, and for that I will be forever thankful.

My mum and dad are just the best—it's a scientific fact, so sorry everyone else, but you're wrong. My mum is marvelously mad, meticulous, and very Mediterranean. My dad is chilled, cool, and hilarious. They defy description. They've brought me nothing but happiness and unending support and care, and I will never be able to adequately repay them. If I'm a tenth as wonderful as they are when all is said and done, then I shall consider that a life well lived.

Stephanie Jane—what can I say? You're a phenomenon. I cannot believe I get to live this ridiculous life with the smartest, goofiest, most adventurous collection of molecules this side of the galaxy. I didn't think we'd have to ride out an actual global disaster together, but I am impossibly lucky to be fending off our foes with my best friend. I'm so glad we both think cricket is boring.

And last but not least: Lola, our mental, silly, clever, gorgeous rescue pup. She jumps up onto the nearest chair or table if you say "The floor is lava!" which makes her the best dog in the world.

NOTES

INTRODUCTION

1. Brown, S. et al. "Volcanic Fatalities Database: Analysis of Volcanic Threat with Distance and Victim Classification." *Journal of Applied Volcanology* 6 (2017). doi: https://doi.org/10.1186/s13617-017-0067-4.
2. Kornei, Katherine. "Ancient Rome Was Teetering. Then a Volcano Erupted 6,000 Miles Away." *New York Times* (June 22, 2020). URL: https://www.nytimes.com/2020/06/22/science/rome-caesar-volcano.html.
3. Welch, Craig. "How Volcanoes Caused Violent Uprisings in Cleopatra's Egypt." *National Geographic* (October 17, 2017). URL: https://www.nationalgeographic.com/news/2017/10/volcanoes-Nile-flood-climate-Egypt.
4. Brannen, Peter. *The Ends of the World: Volcanic Apocalypses, Lethal Oceans and Our Quest to Understand Earth's Past Mass Extinctions.* London: Oneworld Publications, 2017.
5. Andrews, Robin George. "Why the White Island Tragedy Won't Stop the Volcano Tourism Boom." *Wired* (December 14, 2019). URL: https://www.wired.co.uk/article/white-island-volcano-new-zealand.
6. Andrews, Robin George. "Why So Many People Choose to Live Near Active Volcanoes." *Gizmodo* (February 18, 2019). URL: https://earther.gizmodo.com/why-so-many-people-choose-to-live-near-active-volcanoes-1832649751.
7. Todde, A. et al. "The 1914 Taisho Eruption of Sakurajima Volcano: Stratigraphy and Dynamics of the Largest Explosive Event in Japan During the Twentieth Century." *Bulletin of Volcanology* 79 (2017). doi: 10.1007/s00445-017-1154-4.
8. Global Volcanism Program. Aira. [Factbox] URL: https://volcano.si.edu/volcano.cfm?vn=282080.

1. THE FOUNTAIN OF FIRE

1. Fisher, R.V. & Heiken, G. "Mt. Pelée, Martinique: May 8 and 20, 1902, Pyroclastic Flows and Surges." *Journal of Volcanology and Geothermal Research* 13, 3–4 (1982). doi: 10.1016/0377-0273(82)90056-7.

2. Thomas, Gordon & Morgan-Witts, Max. *The Day the World Ended: The Mount Pelée Disaster: May 7, 1902*. New York: Open Road Integrated Media, 2014.
3. Dvorak, John. "The Origin of the Hawaiian Volcano Observatory." *Physics Today* (2011). URL: https://physicstoday.scitation.org/doi/10.1063/1.3592003.
4. Jaggar, Thomas Augustus. *My Experiments with Volcanoes*. Hawai'i: Hawaiian Volcano Research Association, 1956.
5. Global Volcanism Program. Kilauea. [Factbox] URL: https://volcano.si.edu/volcano.cfm?vn=332010.
6. Meyer, Robinson. "A Beginner's Guide to Hawaii's Otherworldly Lava." *Atlantic* (May 9, 2018). URL: https://www.theatlantic.com/science/archive/2018/05/how-to-look-at-hawaiis-lava/559988/.
7. Romero, Simon. "Madame Pele, Hawaii's Goddess of Volcanoes, Awes Those Living in Lava's Path." *New York Times* (May 21, 2018). URL: https://www.nytimes.com/2018/05/21/us/pele-hawaii-volcano.html.
8. Hawaiian Volcano Observatory. (2011) "Volcano Watch — Who Is Frank Alvord Perret, and What Is His Connection to Hawaiian Volcanoes?" [Blog post] URL: https://www.usgs.gov/center-news/volcano-watch-who-frank-alvord-perret-and-what-his-connection-hawaiian-volcanoes.
9. Decker, R.W., Wright, T.L. & Stauffer, P.H., eds. (1987) *Volcanism in Hawaii*. U.S. Geological Survey Professional Paper 1350, 1667 pp. URL: https://pubs.usgs.gov/pp/1987/1350.
10. Hawaiian Volcano Observatory. (2011) "Volcano Watch — The Founding of the Hawaiian Volcano Observatory." [Blog post] URL: https://www.usgs.gov/center-news/volcano-watch-founding-hawaiian-volcano-observatory.
11. Jaggar, T.A. "Thermal Gradient of Kilauea Lava Lake." *Journal of the Washington Academy of Sciences* 7, 13 (1917). URL: https://www.jstor.org/stable/24521345?seq=1#metadata_info_tab_contents.
12. USGS. "Establishing the First U.S. Volcano Observatory." [Blog post] URL: https://www.usgs.gov/observatories/hawaiian-volcano-observatory/establishing-first-us-volcano-observatory.
13. USGS. "The Pu'u 'Ō'ō Eruption Lasted 35 Years." [Blog post] URL: https://www.usgs.gov/volcanoes/kilauea/pu-u-eruption-lasted-35-years.
14. Callis, Tom. "Kilauea Eruption Turning 10 — to the Delight of Scientists and Visitors." *Hawaii Tribune-Herald* (March 18, 2018). URL: https://www.hawaiitribune-herald.com/2018/03/18/hawaii-news/kilauea-eruption-turning-10-to-the-delight-of-scientists-and-visitors/.
15. Marsh, Sarah & Levin, Sam. "Kilauea Volcano: Earthquakes Follow Eruption as Hawaii Orders Evacuations." *Guardian* (May 5, 2018). URL: https://www.theguardian.com/us-news/2018/may/04/hawaii-evacuations-ordered-as-kilauea-volcano-erupts.
16. Associated Press. "Hawaii Volcano Destroys Dozens of Homes, Forces Evacuations."

CNBC (May 7, 2018). URL: https://www.cnbc.com/2018/05/07/hawaii-volcano-destroys-dozens-of-homes-forces-evacuations.html.

17. Lincoln, Mileka. "Lava Threatening Several Additional Wells at Puna Geothermal Plant." *Hawaii News Now* (May 27, 2018). URL: https://www.hawaiinewsnow.com/story/38287296/lava-reaches-site-of-puna-geothermal-plant-no-gas-release-detected/.
18. USGS. "Volcanic Gases Can Be Harmful To Health, Vegetation and Infrastructure." [Blog post] URL: https://www.usgs.gov/natural-hazards/volcano-hazards/volcanic-gases.
19. Stone, Maddie. "Now Kilauea's Eruption Is Producing Wild Blue Flames." *Gizmodo* (May 23, 2018). URL: https://earther.gizmodo.com/now-kilaueas-eruption-is-producing-wild-blue-flames-1826266419.
20. USGS. "Volcano Watch—Can Hawaiian Lava Flows Be Diverted?" [Blog post] URL: https://www.usgs.gov/center-news/volcano-watch-can-hawaiian-lava-flows-be-diverted.
21. Andrews, Robin George. "The Army Bombed a Hawaiian Lava Flow. It Didn't Work." *New York Times* (March 12, 2020). URL: https://www.nytimes.com/2020/03/12/science/volcano-bomb-hawaii.html.
22. Barberi, F. et al. "The Control of Lava Flow During the 1991–1992 Eruption of Mt. Etna." *Journal of Volcanology and Geothermal Research* 56, 1-2 (1993). doi: 10.1016/0377-0273(93)90048-V.
23. Andrews, Robin George. "The Mount St. Helens Eruption Was the Volcanic Warning We Needed." *New York Times* (May 18, 2020). URL: https://www.nytimes.com/2020/05/18/science/mt-st-helens-eruption.html.
24. Neal, C.A. et al. "The 2018 Rift Eruption and Summit Collapse of Kīlauea Volcano." *Science* 363, 6425 (2019). doi: 10.1126/science.aav7046.
25. Dvorsky, George. "Nearly Two Dozen People Injured After Lava Bomb Hits Hawaii Tour Boat." *Gizmodo* (July 18, 2018). URL: https://gizmodo.com/nearly-two-dozen-people-injured-after-lava-bomb-hits-ha-1827639279.
26. Associated Press. "Kilauea Eruption Caused Economic Loss in Area Near National Park." *Hawaii News Now* (May 27, 2019). URL: https://www.hawaiinewsnow.com/2019/05/27/kilauea-eruption-caused-m-loss-area-near-national-park/.
27. Andrews, Robin George. "Witnessing the Birth of a Crater Lake Where Lava Just Flowed." *New York Times* (August 7, 2019). URL: https://www.nytimes.com/2019/08/07/science/hawaii-kilauea-volcano-crater-lake.html.
28. Anderson, K.R. et al. "Magma Reservoir Failure and the Onset of Caldera Collapse at Kīlauea Volcano in 2018." *Science* 366, 6470 (2019). doi: 10.1126/science.aaz1822.

II. THE SUPERVOLCANO

1. Gilbert, Samuel. "Not a Mask in Sight': Thousands Flock to Yellowstone as Park Reopens." *Guardian* (May 19, 2020). URL: https://www.theguardian.com/environment/2020/may/19/yellowstone-coronavirus-reopening-grand-teton-covid-19.

2. Library of Congress. Mapping the National Parks. [Factbox] URL: https://www.loc.gov/collections/national-parks-maps/articles-and-essays/yellowstone-the-first-national-park/.
3. National Park Service. Yellowstone National Park Protection Act (1872). [Document] URL: https://www.nps.gov/yell/learn/management/yellowstoneprotectionact1872.htm.
4. National Park Service. Yellowstone—Flight of the Nez Perce. [Factbox] URL: https://www.nps.gov/yell/learn/historyculture/flightnezperce.htm.
5. Wicks, C.W. et al. "Magma Intrusion and Volatile Ascent Beneath Norris Geyser Basin, Yellowstone National Park." *JGR Solid Earth* (2020). doi: 10.1029/2019JB018208.
6. Andrews, Robin George. "It's Warm and Stealthy, and It Killed Yellowstone Trees and Turned Soil Pale." *New York Times* (April 12, 2019). URL: https://www.nytimes.com/2019/04/12/science/yellowstone-volcano-warm-spot.html.
7. Farrell, J. et al. "Tomography from 26 Years of Seismicity Revealing That the Spatial Extent of the Yellowstone Crustal Magma Reservoir Extends Well Beyond the Yellowstone Caldera." *Geophysical Research Letters* 41, 9 (2014). doi: 10.1002/2014GL059588.
8. Huang, H-H. et al. "The Yellowstone Magmatic System from the Mantle Plume to the Upper Crust." *Science* 348, 6236 (2015). doi: 10.1126/science.aaa5648.
9. Wells, R. et al. "Geologic History of Siletzia, a Large Igneous Province in the Oregon and Washington Coast Range: Correlation to the Geomagnetic Polarity Time Scale and Implications for a Long-Lived Yellowstone Hotspot." *Geosphere* 10, 4 (2014). doi: 10.1130/GES01018.1.
10. Kasbohm, J. & Schoene, B. "Rapid Eruption of the Columbia River Flood Basalt and Correlation with the Mid-Miocene Climate Optimum. *Science Advances* 4, 9 (2018). doi: 10.1126/sciadv.aat8223.
11. Castor, S.B. & Henry, C.D. "Lithium-Rich Claystone in the McDermitt Caldera, Nevada, USA: Geologic, Mineralogical, and Geochemical Characteristics and Possible Origin." *Minerals* 10, 1 (2020). doi: 10.3390/min10010068.
12. Henry, C.D. et al. "Geology and Evolution of the McDermitt Caldera, Northern Nevada and Southeastern Oregon, Western USA." *Geosphere* 13, 4 (2017). doi: 10.1130/GES01454.1.
13. Knott, T.R. et al. "Discovery of Two New Super-eruptions from the Yellowstone Hotspot Track (USA): Is the Yellowstone Hotspot Waning?" *Geology* 48, 9 (2020). doi: 10.1130/G47384.1.
14. Smith, R.B. et al. "Geodynamics of the Yellowstone Hotspot and Mantle Plume: Seismic and GPS Imaging, Kinematics, and Mantle Flow." *Journal of Volcanology and Geothermal Research* 188, 1–3 (2009). doi: 10.1016/j.jvolgeores.2009.08.020.
15. Boyd, F.R. "Welded Tuffs and Flows in the Rhyolite Plateau of Yellowstone Park, Wyoming." *GSA Bulletin* 72, 3 (1961). doi: 10.1130/0016-7606(1961)72[387:WTAFIT]2.0.CO;2.
16. Wallenstein, N. et al. "Origin of the Term Nuées Ardentes and the 1580 and 1808

Eruptions on São Jorge Island, Azores." *Journal of Volcanology and Geothermal Research* 358 (2018). doi: 10.1016/j.jvolgeores.2018.03.022.

17. Andrews, Robin George. "Volcanic 'Avalanches' Glide on Air, Boosting Their Deadly Speed." *National Geographic* (April 8, 2019). URL: https://www.nationalgeographic.com/science/2019/04/volcanic-avalanches-pyroclastic-flows-glide-on-air-boosting-deadly-speed/.
18. USGS. "How Do We Know about the Calderas in Yellowstone?" (Blog post) URL: https://www.usgs.gov/center-news/how-do-we-know-about-calderas-yellowstone.
19. Christiansen, R.L. & Blank Jr., H.R. (1972) *Volcanic Stratigraphy of the Quaternary Rhyolite Plateau in Yellowstone National Park*. U.S. Geological Survey Professional Paper 729-B. doi: 10.3133/pp729B.
20. Christiansen, R.L. (2001) *The Quaternary and Pliocene Yellowstone Plateau Volcanic Field of Wyoming, Idaho, and Montana*. U.S. Geological Survey Professional Paper 729-G. URL: https://pubs.usgs.gov/pp/pp729g/.
21. Myers, M.L. et al. "Prolonged Ascent and Episodic Venting of Discrete Magma Batches at the Onset of the Huckleberry Ridge Supereruption, Yellowstone." *Earth and Planetary Science Letters* 451 (2016). doi: 10.1016/j.epsl.2016.07.023.
22. Swallow, E.J. et al. "The Huckleberry Ridge Tuff, Yellowstone: Evacuation of Multiple Magmatic Systems in a Complex Episodic Eruption." *Journal of Petrology* 60, 7 (2019). doi: 10.1093/petrology/egz034.
23. USGS. (2019) "A Personal Commentary: Why I Dislike the Term 'Supervolcano' (and What We Should Be Saying Instead)." [Blog post] URL: https://www.usgs.gov/center-news/a-personal-commentary-why-i-dislike-term-supervolcano-and-what-we-should-be-saying.
24. Newhall, C.G. & Self, S. "The Volcanic Explosivity Index (VEI) an Estimate of Explosive Magnitude for Historical Volcanism." *JGR Oceans* 87, C2 (1982). doi: 10.1029/JC087iC02p01231.
25. Global Volcanism Program. Taupo. [Factbox] URL: https://volcano.si.edu/volcano.cfm?vn=241070.
26. USGS. "What Is a Supervolcano? What Is a Supereruption?" [Blog post] URL: https://www.usgs.gov/faqs/what-a-supervolcano-what-a-supereruption?qt-news_science_products=0#qt-news_science_products.
27. Cooper, Tracey. "When Will the Next Big Eruption Happen?" *Stuff.co.nz* (November 14, 2009). URL: http://www.stuff.co.nz/national/3062442/When-will-the-next-big-eruption-happen.
28. Mastin, L.G., Van Eaton, A.R. & Lowenstern, J.B. "Modeling Ash Fall Distribution from a Yellowstone Supereruption." *Geochemistry, Geophysics, Geosystems* 15, 8 (2014). doi: 10.1002/2014GC005469.
29. Grzelewski, Derek. "The Power of Taupo." *NZ Geo* (November–December 2009). URL: https://www.nzgeo.com/stories/the-power-of-taupo/.
30. Global Volcanism Program. Toba. [Factbox] URL: https://volcano.si.edu/volcano.cfm?vn=261090.

31. Comprehensive Nuclear-Test-Ban Treaty Organization. (n.d.) 6 July 1962: 'Sedan'—Massive Crater, Massive Contamination. [Factbox] URL: https://www.ctbto.org/specials/testing-times/6-july-1962-sedan-massive-crater-massive-contamination.
32. Global Volcanism Program. Yellowstone. [Factbox] URL: https://volcano.si.edu/volcano.cfm?vn=325010.
33. Kettley, Sebastian. "Yellowstone Volcano Eruption Warning: Hundreds of Bison Dead as Fears of Mega Blast Grow." *Daily Express* (April 5, 2018) URL: https://www.express.co.uk/news/science/941485/Yellowstone-volcano-eruption-bison-cull.
34. Ewert, J.W., Diefenbach, A. & Ramsey, D.W. (2018) *2018 Update to the U.S. Geological Survey National Volcanic Threat Assessment.* U.S. Geological Scientific Investigations Report 2018-5140. doi: 10.3133/sir20185140.
35. National Park Service. Bear-Inflicted Human Injuries and Fatalities in Yellowstone. [Factbox] URL: https://www.nps.gov/yell/learn/nature/injuries.htm.
36. Jackson, Amanda. "A Woman Suffers Burns After Illegally Entering Yellowstone National Park, Park Officials Say." *CNN* (May 13, 2020). URL: https://edition.cnn.com/2020/05/12/us/woman-burned-fell-yellowstone-trnd/index.html.
37. Guarino, Ben. "Man Who Dissolved in Boiling Yellowstone Hot Spring Slipped While Checking Temperature to Take Bath." *Washington Post* (November 17, 2016). URL: https://www.washingtonpost.com/news/morning-mix/wp/2016/11/17/man-who-dissolved-in-boiling-yellowstone-hot-spring-slipped-while-checking-temperature-to-take-bath/.
38. Andrews, Robin George. "Why the New Zealand Volcano Eruption Caught the World by Surprise." *National Geographic* (December 9, 2019). URL: https://www.nationalgeographic.com/science/2019/12/why-new-zealand-volcano-eruption-caught-world-by-surprise-white-island-whakaari/.
39. USGS. Hydrothermal Explosions. [Blog post] URL: https://www.usgs.gov/volcanoes/yellowstone/hydrothermal-explosions.
40. USGS. (2019) "60 Years Since the 1959 M7.3 Hebgen Lake Earthquake: Its History and Effects on the Yellowstone Region." [Blog post] URL: https://www.usgs.gov/center-news/60-years-1959-m73-hebgen-lake-earthquake-its-history-and-effects-yellowstone-region.
41. Deligne, N.I. et al. "Investigating the Consequences of Urban Volcanism Using a Scenario Approach I: Development and Application of a Hypothetical Eruption in the Auckland Volcanic Field, New Zealand." *Journal of Volcanology and Geothermal Research* 336 (2017). doi: 10.1016/j.jvolgeores.2017.02.023.
42. Blake, D.M. et al. "Investigating the Consequences of Urban Volcanism Using a Scenario Approach II: Insights into Transportation Network Damage and Functionality." *Journal of Volcanology and Geothermal Research* 340 (2017). doi: 10.1016/j.jvolgeores.2017.04.010.
43. Robock, A. et al. "Did the Toba Volcanic Eruption of ~74k BP Produce Widespread Glaciation?" *JGR Atmospheres* 114, D10 (2009). doi: 10.1029/2008JD011652.
44. Greshko, Michael. "These Ancient Humans Survived a Supervolcano." *National Geo-*

graphic (March 12, 2018). URL: https://www.nationalgeographic.com/news/2018/03/toba-supervolcano-eruption-humans-south-africa-science/.

45. Petraglia, M. et al. "Middle Paleolithic Assemblages from the Indian Subcontinent Before and After the Toba Super-Eruption." *Science* 317, 5834 (2007). doi: 10.1126/science.1141564.
46. Timmreck, C. et al. "Aerosol Size Confines Climate Response to Volcanic Super-eruptions." *Geophysical Research Letters* 37, 24 (2010). doi: 10.1029/2010GL045464.
47. Lane, C.S., Chorn, B.T. & Johnson, T.C. "Ash from the Toba Supereruption in Lake Malawi Shows No Volcanic Winter in East Africa at 75 ka." *Proceedings of the National Academy of Sciences USA* 110, 20 (2013). doi: 10.1073/pnas.1301474110.
48. Clarkson, C. et al. "Human Occupation of Northern India Spans the Toba Super-eruption ~74,000 Years Ago." *Nature Communications* 11 (2020). doi: 10.1038/s41467-020-14668-4.
49. Wilson, C.J.N. et al. "The 26.5 ka Oruanui Eruption, Taupo Volcano, New Zealand: Development, Characteristics and Evacuation of a Large Rhyolitic Magma Body." *Journal of Petrology* 47, 1 (2006). doi: 10.1093/petrology/egi066.

III. THE GREAT INK WELL

1. Klemetti, Erik. "Strangest Magma on Earth: Carbonatites of Ol Doinyo Lengai." *Wired* (March 11, 2014). URL: https://www.wired.com/2014/03/strangest-magma-earth-carbonatites-oldoinyo-lengai/.
2. Dawson, J.B., Keller, J. & Nyamweru, C. "Historic and Recent Eruptive Activity of Oldoinyo Lengai." In: Bell, K. & Keller, J. (eds.), *Carbonatite Volcanism*. IAVCEI Proceedings in Volcanology, 4 (1995). Berlin: Springer. doi: 10.1007/978-3-642-79182-6_2.
3. Dawson, J.B. "Sodium Carbonate Lavas from Oldoinyo Lengai, Tanganyika." *Nature* 195 (1962). doi: 10.1038/1951075a0.
4. Mattsson, H.B. & Vuorinen, J. "Emplacement and Inflation of Natrocarbonatitic Lava Flows During the March-April 2006 Eruption of Oldoinyo Lengai, Tanzania." *Bulletin of Volcanology* 71, 3 (2009). doi: 10.1007/s00445-008-0224-z.
5. Kervyn, M. et al. "Voluminous Lava Flows at Oldoinyo Lengai in 2006: Chronology of Events and Insights into the Shallow Magmatic System." *Bulletin of Volcanology* 70 (2008). doi: 10.1007/s00445-007-0190-x.
6. Pappas, Stephanie. "'Stone Animal' Lake Seen from Space in All Its Crimson Glory." *LiveScience* (May 9, 2017). URL: https://www.livescience.com/59021-stone-animal-lake-natron-space-photo.html.
7. Sherrod, D.R., Magigita, M.M. & Kwelwa, S. U.S. Geological Survey Geologic Map of Oldonyo Lengai (Oldoinyo Lengai) Volcano and Surroundings, Arusha Region, United Republic of Tanzania: Pamphlet, accompanying report 2013-1306 (2013). URL: https://pubs.usgs.gov/of/2013/1306/pdf/ofr2013-1306_pamphlet.pdf.
8. Global Volcanism Program. Ol Doinyo Lengai. [Factbox] URL: https://volcano.si.edu/showreport.cfm?doi=10.5479/si.GVP.BGVN200802-222120.
9. Davies, G.J. "'Hades'—A Remarkable Cave on Oldoinyo Lengai in the East Afri-

can Rift Valley." *International Journal of Speleology* 27B, 1/4 (1998). URL: https://scholarcommons.usf.edu/cgi/viewcontent.cgi?article=1272&context=ijs.
10. Muirhead, J.D. et al. "Displaced Cratonic Mantle Concentrates Deep Carbon During Continental Rifting." *Nature* 582 (2020). doi: 10.1038/s41586-020-2328-3.
11. Lower, Jaime & Frazzetta, Andrea. "Voyagers: Visual Journeys by Six Photographers." *New York Times Magazine* (September 25, 2016). URL: https://www.nytimes.com/interactive/2016/09/25/magazine/the-voyages-issue.html#Ethiopia.
12. Gómez, F. et al. "Ultra-small Microorganisms in the Polyextreme Conditions of the Dallol Volcano, Northern Afar, Ethiopia." *Scientific Reports* 9 (2019). doi: 10.1038/s41598-019-44440-8.
13. Belilla, J. et al. "Hyperdiverse Archaea Near Life Limits at the Polyextreme Geothermal Dallol Area." *Nature Ecology & Evolution* 3 (2019). doi: 10.1038/s41559-019-1005-0.
14. Andrews, Robin George. "They Didn't Find Life in a Hopeless Place." *New York Times* (November 1, 2019). URL: https://www.nytimes.com/2019/11/01/science/extreme-life-aliens.html.
15. Greshko, Michael. "Treasure Trove of Ancient Human Footprints Found Near Volcano." *National Geographic* (October 10, 2016). URL: https://www.nationalgeographic.com/news/2016/10/ancient-human-footprints-africa-volcano-science/.
16. Hublin, J-J. et al. "New Fossils from Jebel Irhoud, Morocco and the Pan-African Origin of *Homo sapiens*." *Nature* 546 (2017). doi: 10.1038/nature22336.
17. Gibbons, Ann. "World's Oldest *Homo sapiens* Fossils Found in Morocco." *ScienceMag* (June 7, 2017). URL: https://www.sciencemag.org/news/2017/06/world-s-oldest-homo-sapiens-fossils-found-morocco.
18. Díaz, Lucía Pérez. "Africa Is Splitting in Two—Here Is Why." *Conversation* (April 7, 2018). URL: https://theconversation.com/africa-is-splitting-in-two-here-is-why-94056.

IV. THE VAULTS OF GLASS

1. Woods Hole Oceanographic Institution. The Challenger Expedition. [Factbox] URL: https://divediscover.whoi.edu/history-of-oceanography/the-challenger-expedition/.
2. Evan Lubofsky—Woods Hole Oceanographic Institution. (2018) "The Discovery of Hydrothermal Vents." [Blog post] URL: https://www.whoi.edu/oceanus/feature/the-discovery-of-hydrothermal-vents/.
3. National Oceanic and Atmospheric Administration/National Ocean Service. What Is a Hydrothermal Vent? [Factbox] URL: https://oceanservice.noaa.gov/facts/vents.html.
4. Fountain, Henry. "ABE, Pioneering Robotic Undersea Explorer, Is Dead at 16." *New York Times* (March 15, 2010). URL: https://www.nytimes.com/2010/03/16/science/16sub.html.
5. Cook, Gareth. "The Lives They Lived." *New York Times* (December 23, 2019). URL: https://www.nytimes.com/interactive/2019/12/23/magazine/opportunity-rover.html.
6. Andrews, Robin George. "A Vault of Glass and the Deepest Volcanic Eruption

Ever Detected." *New York Times* (October 30, 2018). URL: https://www.nytimes.com/2018/10/30/science/deep-sea-volcano.html.

7. Andrews, Robin George. "Volcano Space Robots Are Prepping for a Wild Mission to Jupiter." *Wired* (December 13, 2019). URL: https://www.wired.co.uk/article/nasa-submarines-searching.
8. Schnur, S.R. et al. "A Decade of Volcanic Construction and Destruction at the Summit of NW Rota-1 Seamount: 2004–2014." *JGR Solid Earth* 122, 3 (2017). doi: 10.1002/2016JB013742.
9. Embley, R.W. et al. "Eruptive Modes and Hiatus of Volcanism at West Mata Seamount, NE Lau Basin: 1996–2012." *Geochemistry, Geophysics, Geosystems* 15, 10 (2014). doi: 10.1002/2014GC005387.
10. Columbia University Earth Institute (2020). "Marie Tharp's Adventures in Mapping the Seafloor, In Her Own Words." [Blog post] URL: https://blogs.ei.columbia.edu/2020/07/24/marie-tharp-connecting-dots/.
11. National Geographic. Continental Drift. [Factbox] URL: https://www.nationalgeographic.org/encyclopedia/continental-drift/.
12. Andrews, Robin George. "A Deep-Sea Magma Monster Gets a Body Scan." *New York Times* (December 3, 2019). URL: https://www.nytimes.com/2019/12/03/science/axial-volcano-mapping.html.
13. Andrews, Robin George. "Coronavirus Turns Urban Life's Roar to Whisper on World's Seismographs." *New York Times* (April 8, 2020). URL: https://www.nytimes.com/2020/04/08/science/seismographs-lockdown-coronavirus.html.
14. Andrews, Robin George. "Geologists Joke About 'Sea Monster' After Mysterious 30-Minute Rumble Emanates from Waters Near Madagascar." *Gizmodo* (November 29, 2018). URL: https://gizmodo.com/geologists-joke-about-sea-monster-after-mysterious-30-m-1830738921.
15. Wei-Haas, Maya. "Strange Waves Rippled Around the World, and Nobody Knows Why." *National Geographic* (November 28, 2018). URL: https://www.nationalgeographic.com/science/2018/11/strange-earthquake-waves-rippled-around-world-earth-geology/.
16. Andrews, Robin George. "Scientists Witness the Birth of a Submarine Volcano for the First Time." *Gizmodo* (May 24, 2019). URL: https://gizmodo.com/scientists-witness-the-birth-of-a-submarine-volcano-for-1834990629.
17. Vine, F.J. & Matthews, D.H. "Magnetic Anomalies Over Oceanic Ridges." *Nature* 199 (1963). doi: 10.1038/199947a0.
18. OConnell, Suzanne. "Marie Tharp's Maps Revolutionized Our Knowledge of the Seafloor." *Washington Post* (August 8, 2020). URL: https://www.washingtonpost.com/science/marie-tharps-maps-revolutionized-our-knowledge-of-the-seafloor/2020/08/07/d9d112bc-d767-11ea-9c3b-dfc394c03988_story.html.
19. Andrews, Robin George. "Gigantic Pumice Raft from Underwater Eruption Is on a Wild Ride Across the Pacific Ocean." *Gizmodo* (August 26, 2019). URL:

https://gizmodo.com/gigantic-pumice-raft-from-underwater-eruption-is-on-a-w-1837581054.

20. Dürig, T. et al. "Deep-Sea Eruptions Boosted by Induced Fuel–Coolant Explosions." *Nature Geoscience* 13 (2020). doi: 10.1038/s41561-020-0603-4.
21. Sturdy, E.W. "The Volcanic Eruption of Krakatoa." *Atlantic* (September, 1884). URL: https://www.theatlantic.com/magazine/archive/1884/09/the-volcanic-eruption-of-krakatoa/376174/.
22. Hernandez, Marco & Scarr, Simon. "How Powerful Was the Beirut Blast?" *Reuters* (August 14, 2020). URL: https://graphics.reuters.com/LEBANON-SECURITY/BLAST/yzdpxnmqbpx/.
23. Bhatia, Aatish. "The Loudest Sound Ever Heard." *Discover* (July 13, 2018). URL: https://www.discovermagazine.com/environment/the-loudest-sound-ever-heard.
24. National Centers for Environmental Information. "On This Day: Historic Krakatau Eruption of 1883." [Blog post] URL: https://www.ncei.noaa.gov/news/day-historic-krakatau-eruption-1883.
25. Amos, Jonathan. "Anak Krakatau: Volcano's Tsunami Trigger Was 'Relatively Small.'" *BBC News* (September 3, 2019). URL: https://www.bbc.co.uk/news/science-environment-49568107.
26. Williams, R., Rowley, P. & Garthwaite, M.C. "Reconstructing the Anak Krakatau Flank Collapse That Caused the December 2018 Indonesian Tsunami." *Geology* 47, 10 (2019). doi: 10.1130/G46517.1
27. Salinas-de-León, P. et al. "Deep-Sea Hydrothermal Vents as Natural Egg-Case Incubators at the Galapagos Rift." *Scientific Reports* 8 (2018). doi: 10.1038/s41598-018-20046-4.
28. Andrews, Robin George. "Bizarre Life-forms Found Thriving in Ancient Rocks Beneath the Seafloor." *National Geographic* (April 2, 2020). URL: https://www.nationalgeographic.com/science/2020/04/life-found-thriving-in-one-of-the-least-likely-spots-on-earth/.
29. Wu, Katherine J. "These Microbes May Have Survived 100 Million Years Beneath the Seafloor." *New York Times* (July 28, 2020). URL: https://www.nytimes.com/2020/07/28/science/microbes-100-million-years-old.html.

V. THE PALE GUARDIAN

1. USGS. Unified Geologic Map of the Moon, 1:5M, 2020. [Document] URL: https://astrogeology.usgs.gov/search/map/Moon/Geology/Unified_Geologic_Map_of_the_Moon_GIS.
2. Cartier, Kimberly M.S. "Apollo May Have Found an Earth Meteorite on the Moon." *Eos* (January 28, 2019). URL: https://eos.org/articles/apollo-may-have-found-an-earth-meteorite-on-the-moon.
3. Andrews, Robin George. "During the Lunar Eclipse, Something Slammed Into the Moon." *New York Times* (January 23, 2019). URL: https://www.nytimes.com/2019/01/23/science/lunar-eclipse-meteor-moon.html.

4. Boyle, Rebecca. "What Made the Moon? New Ideas Try to Rescue a Troubled Theory." *Quanta Magazine* (August 2, 2017). URL: https://www.quantamagazine.org/what-made-the-moon-new-ideas-try-to-rescue-a-troubled-theory-20170802/.
5. Melosh, H.J. "Why the Moon Is so Like the Earth." *Nature Geoscience* 12 (2019). doi: 10.1038/s41561-019-0364-0.
6. Hosono, N. et al. "Terrestrial Magma Ocean Origin of the Moon." *Nature Geoscience* 12 (2019). doi: 10.1038/s41561-019-0354-2.
7. Snape, J. et al. "Ancient Volcanism on the Moon: Insights from Pb Isotopes in the MIL 13317 and Kalahari 009 Lunar Meteorites." *Earth and Planetary Science Letters* 502 (2018). doi: 10.1016/j.epsl.2018.08.035.
8. Verne, Jules. *From the Earth to the Moon* (Illustrated 1874 Edition: 100th Anniversary Collection). Orinda, CA: SeaWolf Press, 2019.
9. Aderin-Pocock, Maggie. "A Guide to the Moon's Craters, Seas, and Ghostly Shine." *Popular Science* (April 9, 2019). URL: https://www.popsci.com/book-of-the-moon-excerpt/.
10. Sigurdsson, H. et al., eds. *The Encyclopedia of Volcanoes*, 2nd ed. London: Academic Press/Elsevier, 2015
11. Elkins-Tanton, L.T., Hager, B.H. & Grove, T.L. "Magmatic Effects of the Lunar Late Heavy Bombardment." *Earth and Planetary Science Letters* 22, 1 (2004). doi: 10.1016/j.epsl.2004.02.017.
12. Lakdawalla, Emily—The Planetary Society (2011). "The Moon is a KREEPy place." [Blog post] URL: https://www.planetary.org/articles/3013.
13. Pasckert, J.H., Hiesinger, H. & van der Bogert, C.H. "Small-Scale Lunar Farside Volcanism." *Icarus* 257 (2015). doi: 10.1016/j.icarus.2015.04.040.
14. Roy, A., Wright, J.T. & Sigursson, S. "Earthshine on a Young Moon: Explaining the Lunar Farside Highlands." *Astrophysical Journal Letters* 788, 2 (2014). doi: 10.1088/2041-8205/788/2/L42.
15. Zhu, M-H. et al. "Are the Moon's Nearside-Farside Asymmetries the Result of a Giant Impact?" *JGR Planets* 124, 1 (2019). doi: 10.1029/2018JE005826.
16. Potter, R.W.K. et al. "Constraining the Size of the South Pole-Aitken Basin Impact." *Icarus* 220, 2 (2012). doi: 10.1016/j.icarus.2012.05.032.
17. Wieczorek, M.A., Weiss, B.P. & Stewart, S.T. "An Impactor Origin for Lunar Magnetic Anomalies." *Science* 335, 6073 (2012). doi: 10.1126/science.1214773.
18. Sruthi, U. & Kumar, P.S. "Volcanism on Farside of the Moon: New Evidence from Antoniadi in South Pole Aitken Basin." *Icarus* 242 (2014). doi: 10.1016/j.icarus.2014.07.030.
19. Pasckert, J.H., Hiesinger, H. & van der Bogert, C.H. "Lunar Farside Volcanism in and Around the South Pole–Aitken Basin. *Icarus* 299 (2018). doi: 10.1016/j.icarus.2017.07.023.
20. Andrews-Hanna, J.C. et al. "Structure and Evolution of the Lunar Procellarum Region as Revealed by GRAIL Gravity Data." *Nature* 514 (2014). doi: 10.1038/nature13697.
21. Heiken, G.H., McKay, D.S. & Brown, R.W. "Lunar Deposits of Possi-

ble Pyroclastic Origin." *Geochimica et Cosmochimica Acta* 38, 11 (1974). doi: 10.1016/0016-7037(74)90187-2.

22. Needham, D.H. & Kring, D.A. "Lunar Volcanism Produced a Transient Atmosphere Around the Ancient Moon." *Earth and Planetary Science Letters* 478 (2017). doi: 10.1016/j.epsl.2017.09.002.
23. Brown, David W. "NASA Considers a Rover Mission to Go Cave Diving on the Moon." *Smithsonian* (March 26, 2019). URL: https://www.smithsonianmag.com/science-nature/nasa-considers-rover-mission-go-cave-diving-moon-180971790/.
24. Garry, W.B. et al. "The Origin of Ina: Evidence for Inflated Lava Flows on the Moon." *JGR Planets* 117, 12 (2012). doi: 10.1029/2011JE003981.
25. Qiao, L. et al. "Ina Pit Crater on the Moon: Extrusion of Waning-Stage Lava Lake Magmatic Foam Results in Extremely Young Crater Retention Ages." *Geology* 45, 5 (2017). doi: 10.1130/G38594.1.
26. Braden, S.E. et al. "Evidence for Basaltic Volcanism on the Moon within the past 100 Million Years." *Nature Geoscience* 7 (2014). doi: 10.1038/ngeo2252.
27. Andrews, Robin. "The Moon Is a Volcanic Freak and China Is Trying to Find Out Why." *Wired* (December 08, 2020). URL: https://www.wired.co.uk/article/china-moon-mission-change-5.
28. Sori, M.M. et al. "Gravitational Search for Cryptovolcanism on the Moon: Evidence for Large Volumes of Early Igneous Activity." *Icarus* 273 (2016). doi: 10.1016/j.icarus.2016.02.009.
29. Peters, K. & Langseth, M.G. (1975) *Long Term Temperature Observations on the Lunar Surface at Apollo Sites 15 and 17*. Technical Report 3-CU-3-75, Lamont–Doherty Geological Observatory of Columbia University. URL: https://www.lpi.usra.edu/lunar/ALSEP/pdf/31111000591808.pdf.
30. Nagihara, S. et al. "Examination of the Long-Term Subsurface Warming Observed at the Apollo 15 and 17 Sites Utilizing the Newly Restored Heat Flow Experiment Data From 1975 to 1977." *JGR Planets* 123, 5 (2018). doi: 10.1029/2018JE005579.
31. Watters, T.R. et al. "Shallow Seismic Activity and Young Thrust Faults on the Moon." *Nature Geoscience* 12 (2019). doi: 10.1038/s41561-019-0362-2.
32. Saal, A.E. et al. "Volatile Content of Lunar Volcanic Glasses and the Presence of Water in the Moon's Interior." *Nature* 454 (2008). doi: 10.1038/nature07047.
33. Milliken, R.E. & Li, S. "Remote Detection of Widespread Indigenous Water in Lunar Pyroclastic Deposits." *Nature Geoscience* 10 (2017). doi: 10.1038/ngeo2993.
34. Barnes, J.J. et al. "An Asteroidal Origin for Water in the Moon." *Nature Communications* 7 (2016). doi: 10.1038/ncomms11684.

VI. THE TOPPLED GOD

1. Howell, Elizabeth. "Mariner 9: First Spacecraft to Orbit Mars." *Space.com* (November 08, 2018). URL: https://www.space.com/18439-mariner-9.html.
2. Andrews-Hanna, J.C., Zuber, M.T. & Banerdt, W.B. "The Borealis Basin and

the Origin of the Martian Crustal Dichotomy." *Nature* 453 (2008). doi: 10.1038/nature07011.

3. Leone, G. et al. "Three-Dimensional Simulations of the Southern Polar Giant Impact Hypothesis for the Origin of the Martian Dichotomy." *Geophysical Research Letters* 41, 24 (2014). doi: 10.1002/2014GL062261.
4. Wilson, L. & Mouginis-Mark, P.J. "Phreatomagmatic Explosive Origin of Hrad Vallis, Mars." *JGR Planets* 108, E8 (2003). doi: 10.1029/2002JE001927.
5. Brož, P. & Hauber, E. "Hydrovolcanic Tuff Rings and Cones as Indicators for Phreatomagmatic Explosive Eruptions on Mars." *JGR Planets* 118, 8 (2013). doi: 10.1002/jgre.20120.
6. Wall, K.T. et al. "Determining Volcanic Eruption Styles on Earth and Mars from Crystallinity Measurements." *Nature Communications* 5 (2014). doi: 10.1038/ncomms6090.
7. Jet Propulsion Laboratory. (2016) Found: Clues about Volcanoes Under Ice on Ancient Mars. [Press release] URL: https://www.jpl.nasa.gov/news/news.php?feature=6472.
8. Brož, P. et al. "Experimental Evidence for Lava-Like Mud Flows Under Martian Surface Conditions." *Nature Geoscience* 13 (2020). doi: 10.1038/s41561-020-0577-2.
9. NASA. (2018) Jamming with the "Spiders" from Mars. [Press release] URL: https://www.nasa.gov/image-feature/jpl/jamming-with-the-spiders-from-mars.
10. The Planetary Society. Every Mission to Mars, Ever. [Factbox] URL: https://www.planetary.org/space-missions/every-mars-mission.
11. Andrews, Robin George. "Rocks, Rockets and Robots: The Plan to Bring Mars Down to Earth." *Scientific American* (February 6, 2020). URL: https://www.scientificamerican.com/article/rocks-rockets-and-robots-the-plan-to-bring-mars-down-to-earth1/.
12. International Meteorite Collectors Association. Martian Meteorites. [Factbox] URL: https://imca.cc/mars/martian-meteorites.htm.
13. Lapen, T.J. et al. "Two Billion Years of Magmatism Recorded from a Single Mars Meteorite Ejection Site." *Geology* 3, 2 (2017). doi: 10.1126/sciadv.1600922.
14. Carr, M.H. & Head III, J.W. "Geologic History of Mars." *Earth and Planetary Science Letters* 294 (2010). doi: 10.1016/j.epsl.2009.06.042.
15. Bouley, S. et al. "The Revised Tectonic History of Tharsis." *Earth and Planetary Science Letters* 488 (2018). doi: 10.1016/j.epsl.2018.02.019.
16. Lillis, R.J. et al. "Demagnetization of Crust by Magmatic Intrusion Near the Arsia Mons Volcano: Magnetic and Thermal Implications for the Development of the Tharsis Province, Mars." *Journal of Volcanology and Geothermal Research* 185, 1–2 (2009). doi: 10.1016/j.jvolgeores.2008.12.007.
17. Bouley, S. et al. "Late Tharsis Formation and Implications for Early Mars." *Nature* 531 (2016). doi: 10.1038/nature17171.
18. Ojha, L. et al. "Spectral Evidence for Hydrated Salts in Recurring Slope Lineae on Mars." *Nature Geoscience* 8 (2015). doi: 10.1038/ngeo2546.

19. DiBiase, R.A. et al. "Deltaic Deposits at Aeolis Dorsa: Sedimentary Evidence for a Standing Body of Water on the Northern Plains of Mars." *JGR Planets* 118, 6 (2013). doi: 10.1002/jgre.20100.
20. Bibring, J-P. et al. "Global Mineralogical and Aqueous Mars History Derived from OMEGA/Mars Express Data." *Science* 312, 5772 (2006). doi: 10.1126/science.1122659.
21. Murchie, S.L. et al. "A Synthesis of Martian Aqueous Mineralogy After 1 Mars Year of Observations from the Mars Reconnaissance Orbiter." *JGR Planets* 114, E2 (2009). doi: 10.1029/2009JE003342.
22. Jet Propulsion Laboratory. (2015) NASA's Curiosity Rover Team Confirms Ancient Lakes on Mars. [Press release] URL: https://www.jpl.nasa.gov/news/news.php?feature=4734.
23. Rapin, W. et al. "An Interval of High Salinity in Ancient Gale Crater Lake on Mars." *Nature Geoscience* 12 (2019). doi: 10.1038/s41561-019-0458-8.
24. Villanueva, G.L. et al. "Strong Water Isotopic Anomalies in the Martian Atmosphere: Probing Current and Ancient Reservoirs." *Science* 348, 6231 (2015). doi: 10.1126/science.aaa3630.
25. Rodriguez, J.A.P. et al. "The 1997 Mars Pathfinder Spacecraft Landing Site: Spillover Deposits from an Early Mars Inland Sea." *Scientific Reports* 9 (2019). doi: 10.1038/s41598-019-39632-1.
26. Rodriguez, J.A.P. et al. "Tsunami Waves Extensively Resurfaced the Shorelines of an Early Martian Ocean." *Scientific Reports* 6 (2016). doi: 10.1038/srep25106.
27. Mahaffy, P.R. et al. "Abundance and Isotopic Composition of Gases in the Martian Atmosphere from the Curiosity Rover." *Science* 341, 6143 (2013). doi: 10.1126/science.1237966.
28. Webster, C.R. et al. "Isotope Ratios of H, C, and O in CO_2 and H_2O of the Martian Atmosphere." *Science* 341, 6143 (2013). doi: 10.1126/science.1237961.
29. Turbet, M. & Forget, F. "The Paradoxes of the Late Hesperian Mars Ocean." *Scientific Reports* 9 (2019). doi: 10.1038/s41598-019-42030-2.
30. Lopes, R.M.C., Guest, J.E. & Wilson, C.J. "Origin of the Olympus Mons Aureole and Perimeter Scarp." *The Moon and the Planets* 22 (1980). doi: 10.1007/BF00898433.
31. Isherwood, R.J. et al. "The Volcanic History of Olympus Mons from Paleotopography." *Earth and Planetary Science Letters* 363 (2013). doi: 10.1016/j.epsl.2012.12.020.
32. Richardson, J.A. et al. "Recurrence Rate and Magma Effusion Rate for the Latest Volcanism on Arsia Mons, Mars." *Earth and Planetary Science Letters* 458 (2017). doi: 10.1016/j.epsl.2016.10.040.
33. Wei-Hass, Maya. "First Active Fault Zone Found on Mars." *National Geographic* (December 24, 2019). URL: https://www.nationalgeographic.com/science/2019/12/first-active-fault-system-found-mars2/.

VII. THE INFERNO

1. NASA. Venus. [Factbox] URL: https://solarsystem.nasa.gov/planets/venus/by-the-numbers/.
2. NASA. (2017) "55 Years Ago: Mariner 2 First to Venus." [Blog post] URL: https://www.nasa.gov/feature/55-years-ago-mariner-2-first-to-venus.
3. NASA. Venera 4. [Factbox] URL: https://solarsystem.nasa.gov/missions/venera-4/in-depth/.
4. Guinness World Records. Longest Time Survived on Venus by a Spacecraft. [Factbox] URL: https://www.guinnessworldrecords.com/world-records/78367-longest-time-survived-on-venus-by-a-spacecraft/.
5. Mouginis-Mark, P.J. "Geomorphology and Volcanology of Maat Mons, Venus." *Icarus* 277 (2016). doi: 10.1016/j.icarus.2016.05.022.
6. Krassilnikov, A.S. & Head, J.W. "Novae on Venus: Geology, Classification, and Evolution." *JGR Planets* 108, E9 (2003). doi: 10.1029/2002JE001983.
7. Wilford, John Noble. "On Venus, Pancakes for Volcano Domes." *New York Times* (November 17, 1990). URL: https://www.nytimes.com/1990/11/17/us/on-venus-pancakes-for-volcano-domes.html.
8. Wilford, John Noble. "New Images Suggest Volcanism On Venus." *New York Times* (November 22, 1983). URL: https://www.nytimes.com/1983/11/22/science/new-images-suggest-volcanism-on-venus.html.
9. Smrekar, S.E. et al. "Recent Hotspot Volcanism on Venus from VIRTIS Emissivity Data." *Science* 328, 5978 (2010). doi: 10.1126/science.1186785.
10. Hall, Shannon. "Volcanoes on Venus Might Still Be Smoking." *New York Times* (January 9, 2020). URL: https://www.nytimes.com/2020/01/09/science/venus-volcanoes-active.html.
11. Filiberto, J. et al. "Present-Day Volcanism on Venus as Evidenced from Weathering Rates of Olivine." *Science Advances* 6, 1 (2020). doi: 10.1126/sciadv.aax7445.
12. Marcq, E. et al. "Variations of Sulphur Dioxide at the Cloud Top of Venus's Dynamic Atmosphere." *Nature Geoscience* 6 (2013). doi: 10.1038/ngeo1650.
13. Gülcher, A.J.P. et al. "Corona Structures Driven by Plume–Lithosphere Interactions and Evidence for Ongoing Plume Activity on Venus." *Nature Geoscience* 13 (2020). doi: 10.1038/s41561-020-0606-1.
14. Way, M.J. & Genio, A.D.D. "Venusian Habitable Climate Scenarios: Modeling Venus Through Time and Applications to Slowly Rotating Venus-Like Exoplanets." *JGR Planets* 125, 5 (2020). doi: 10.1029/2019JE006276.
15. Elkins-Tanton, L.T. et al. "Field Evidence for Coal Combustion Links the 252 Ma Siberian Traps with Global Carbon Disruption." *Geology* 48, 10 (2020). doi: 10.1130/G47365.1.
16. Drake, Nadia. "Possible Sign of Life on Venus Stirs Up Heated Debate." *National Geographic* (September 14, 2020). URL: https://www.nationalgeographic.com/science/2020/09/possible-sign-of-life-found-on-venus-phosphine-gas/.
17. Scoles, Sarah. "'Dr. Phosphine' and the Possibility of Life on Venus." *Wired* (Septem-

ber 14, 2020). URL: https://www.wired.com/story/dr-phosphine-and-the-possibility-of-life-on-venus/.

18. Sousa-Silva, C. et al. "Phosphine as a Biosignature Gas in Exoplanet Atmospheres." *arXiv* (2019). doi: 10.1089/ast.2018.1954.
19. Greaves, J.S. et al. "Phosphine Gas in the Cloud Decks of Venus." *Nature Astronomy* (2020a). doi: 10.1038/s41550-020-1174-4.
20. Stirone, Shannon, Chang, Kenneth & Overbye, Dennis. "Life on Venus? Astronomers See a Signal in Its Clouds." *New York Times* (September 14, 2020). URL: https://www.nytimes.com/2020/09/14/science/venus-life-clouds.html.
21. Morowitz, H. & Sagan, C. "Life in the Clouds of Venus?" *Nature* 215 (1967). doi: 10.1038/215159a0.
22. Limaye, S.S. et al. "Venus' Spectral Signatures and the Potential for Life in the Clouds." *Astrobiology* 18, 9 (2018). doi: 10.1089/ast.2017.1783.
23. Wall, K. et al. "Biodiversity Hot Spot on a Hot Spot: Novel Extremophile Diversity in Hawaiian Fumaroles." *MicrobiologyOpen* 4, 2 (2015). doi: 10.1002/mbo3.236.
24. Makuch-Schulze, D. et al. "A Sulfur-Based Survival Strategy for Putative Phototrophic Life in the Venusian Atmosphere." *Astrobiology* 4, 1 (2004). doi: 10.1089/153110704773600203.
25. Seager, S. et al. "The Venusian Lower Atmosphere Haze as a Depot for Desiccated Microbial Life: A Proposed Life Cycle for Persistence of the Venusian Aerial Biosphere." *Astrobiology* (2020). doi: 10.1089/ast.2020.2244.
26. Villanueva, G. et al. "No Phosphine in the Atmosphere of Venus." *arXiv* (2020). URL: https://arxiv.org/abs/2010.14305.
27. Snellen, I.A.G. et al. "Re-analysis of the 267-GHz ALMA Observations of Venus: No Statistically Significant Detection of Phosphine." *arXiv* (2020). URL: https://arxiv.org/abs/2010.09761.
28. Mogol, R., Limaye, S. & Way, M. "Venus' Mass Spectra Show Signs of Disequilibria in the Middle Clouds." *ESSOAr* (2020). doi: 10.1002/essoar.10504552.1.
29. Greaves, J.S. et al. "Re-analysis of Phosphine in Venus' Clouds." *arXiv* (2020b). URL: https://arxiv.org/abs/2011.08176.
30. Andrews, Robin George. "Burying CAESAR: How NASA Picks Winners—and Losers—in Space Exploration." *Scientific American* (July 25, 2019). URL: https://www.scientificamerican.com/article/burying-caesar-how-nasa-picks-winners-and-losers-in-space-exploration/.
31. Odynova, Alexandra. "Russia's Space Agency Chief Declares Venus a 'Russian Planet.'" *CBS News* (September 17, 2020). URL: https://www.cbsnews.com/news/venus-russian-planet-space-agency-chief-claims/.
32. Choi, Charles Q. "Mars Life? 20 Years Later, Debate Over Meteorite Continues." *Space.com* (August 10, 2016). URL: https://www.space.com/33690-allen-hills-mars-meteorite-alien-life-20-years.html.

VIII. THE GIANT'S FORGE

1. Jet Propulsion Laboratory. Voyager Mission Timeline. [Factbox] URL: https://voyager.jpl.nasa.gov/mission/timeline/#event-a-once-in-a-lifetime-alignment.
2. Andrews, Robin George. "Uranus Ejected a Giant Plasma Bubble During Voyager 2's Visit." *New York Times* (March 27, 2020). URL: https://www.nytimes.com/2020/03/27/science/uranus-bubble-voyager.html.
3. Jet Propulsion Laboratory. Did You Know?—Voyager. [Factbox] URL: https://voyager.jpl.nasa.gov/mission/did-you-know/.
4. Morabito, L.A. "Discovery of Volcanic Activity on Io—A Historical Review." *arXiv* (2012). URL: https://arxiv.org/pdf/1211.2554.pdf.
5. NASA. Jupiter Moons. [Factbox] URL: https://solarsystem.nasa.gov/moons/jupiter-moons/in-depth/.
6. Starr, Michelle. "You'll Never Guess What Scientists Want to Call the Moon of a Moon." *ScienceAlert* (October 11, 2018). URL: https://www.sciencealert.com/can-a-moon-have-a-moon-kollmeister-raymond-submoon-moonmoon.
7. Peale, S.J., Cassen, P. & Reynolds, R.T. "Melting of Io by Tidal Dissipation." *Science* 203, 4383 (1979). doi: 10.1126/science.203.4383.892.
8. Andrews, Robin George. "A Dance That Stops 2 of Neptune's Moons from Colliding." *New York Times* (November 21, 2019). URL: https://www.nytimes.com/2019/11/21/science/neptune-moons-orbit.html.
9. Andrews, Robin George. "Are Saturn's Rings Really as Young as the Dinosaurs?" *Quanta Magazine* (November 21, 2019). URL: https://www.quantamagazine.org/are-saturns-rings-really-as-young-as-the-dinosaurs-20191121/.
10. The Planetary Society. Orbital Resonances of the Galilean Moons of Jupiter. [Factbox] URL: https://www.planetary.org/space-images/orbital-resonances-of-galilean-moons.
11. Morabito, L.A. et al. "Discovery of Currently Active Extraterrestrial Volcanism." *Science* 204, 4396 (1979). doi: 10.1126/science.204.4396.972.
12. Jet Propulsion Laboratory. Mission Status—Voyager. [Factbox] URL: https://voyager.jpl.nasa.gov/mission/status/.
13. NASA. Galileo. [Factbox] URL: https://solarsystem.nasa.gov/missions/galileo/overview/.
14. Khurana, K.K. et al. "Evidence of a Global Magma Ocean in Io's Interior." *Science* 332, 6034 (2011). doi: 10.1126/science.1201425.
15. Davies, A.G. "Volcanism on Io: The View from Galileo." *Astronomy & Geophysics* 42, 2 (2001). doi: 10.1046/j.1468-4004.2001.42210.x.
16. de Kleer, K. et al. "Multi-phase Volcanic Resurfacing at Loki Patera on Io." *Nature* 545 (2017). doi: 10.1038/nature22339.
17. Johns Hopkins University Applied Physics Laboratory—New Horizons (2007). "Tvashtar's Plume." [Blog post] URL: http://pluto.jhuapl.edu/Galleries/Featured-Images/image.php?page=6&gallery_id=2&image_id=31.

18. Tsang, C.C.C. et al. "The Collapse of Io's Primary Atmosphere in Jupiter Eclipse." *JGR Planets* 121, 8 (2016). doi: 10.1002/2016JE005025.
19. Guinness World Records. Most Powerful Volcanic Eruption Recorded in the Solar System. [Factbox] URL: https://www.guinnessworldrecords.com/world-records/most-powerful-volcanic-eruption-recorded-in-the-solar-system.
20. Williams, D.A. et al. "The Summer 1997 Eruption at Pillan Patera on Io: Implications for Ultrabasic Lava Flow Emplacement." *JGR Planets* 106, E12 (2001). doi: 10.1029/2000JE001339.
21. Hamilton, C.W. et al. "Spatial Distribution of Volcanoes on Io: Implications for Tidal Heating and Magma Ascent." *Earth and Planetary Science Letters* 361 (2013). doi: 10.1016/j.epsl.2012.10.032.
22. Davies, A.G., Keszthelyi, L.P. & McEwen, A.S. "Determination of Eruption Temperature of Io's Lavas Using Lava Tube Skylights." *Icarus* 278 (2016). doi: 10.1016/j.icarus.2016.06.003.
23. Stirone, Shannon. "Dear Cassini: Why the Saturn Spacecraft Brings Me to Tears." *National Geographic* (September 14, 2018). URL: https://www.nationalgeographic.com/science/2018/09/news-cassini-grand-tour-finale-nasa-space/.
24. NASA. Cassini—Enceladus: Ocean Moon. [Factbox] URL: https://solarsystem.nasa.gov/missions/cassini/science/enceladus/.
25. NASA. Ceres. [Factbox] URL: https://solarsystem.nasa.gov/planets/dwarf-planets/ceres/overview/.
26. Ruesch, O. et al. "Cryovolcanism on Ceres." *Science* 353, 6303 (2016). doi: 10.1126/science.aaf4286.
27. Sori, M.M. et al. "Cryovolcanic Rates on Ceres Revealed by Topography." *Nature Astronomy* 2 (2018). doi: 10.1038/s41550-018-0574-1.
28. Lewis, J.S. "Satellites of the Outer Planets: Their Physical and Chemical Nature." *Icarus* 15, 2 (1971). doi: 10.1016/0019-1035(71)90072-8.
29. Purdue University. An Ammonia-Water Slurry May Swirl Below Pluto's Icy Surface. [Press release] URL: https://www.purdue.edu/newsroom/releases/2015/Q4/an-ammonia-water-slurry-may-swirl-below-plutos-icy-surface.html.
30. Kamata, S. et al. "Pluto's Ocean Is Capped and Insulated by Gas Hydrates." *Nature Geoscience* 12 (2019). doi: 10.1038/s41561-019-0369-8.
31. Dawn XM2 at Occator Crater—Paper Collection. *Nature Astronomy* 11 (2020). URL: https://www.nature.com/collections/agdgfadcag.
32. Andrews, Robin George. "Collision on One Side of Pluto Ripped Up Terrain on the Other, Study Suggests." *Scientific American* (March 26, 2020). URL: https://www.scientificamerican.com/article/collision-on-one-side-of-pluto-ripped-up-terrain-on-the-other-study-suggests/.
33. Hershberger, Scott. "Claims of 'Ocean' Inside Ceres May Not Hold Water." *Scientific American* (August 12, 2020). URL: https://www.scientificamerican.com/article/claims-of-ocean-inside-ceres-may-not-hold-water/.

34. Luger, R., Sestovic, M. & Queloz, D. "A Seven-Planet Resonant Chain in TRAPPIST-1." *Nature Astronomy* 1 (2017). doi: 10.1038/s41550-017-0129.

EPILOGUE: THE TIME TRAVELER

1. Mastrolorenzo, G. et al. "Herculaneum Victims of Vesuvius in AD 79." *Nature* 410 (2001). doi: 10.1038/35071167.
2. Mastrolorenzo, G. et al. "Lethal Thermal Impact at Periphery of Pyroclastic Surges: Evidences at Pompeii." *PLOS ONE* 5, 6 (2010). doi: 10.1371/journal.pone.0011127.
3. Petrone, P. et al. "A Hypothesis of Sudden Body Fluid Vaporization in the 79 AD Victims of Vesuvius." *PLOS ONE* 13, 9 (2018). doi: 10.1371/journal.pone.0203210.
4. Petrone, P. et al. "Heat-Induced Brain Vitrification from the Vesuvius Eruption in C.E. 79." *New England Journal of Medicine* 382 (2020). doi: 10.1056/NEJMc1909867.
5. Andrews, Robin George. "Did Vesuvius Vaporize Its Victims? Get the Facts." *National Geographic* (October 16, 2018). URL: https://www.nationalgeographic.com/science/2018/10/news-pompeii-deaths-vesuvius-vaporized-skulls-exploded-chemistry/.
6. Andrews, Robin George. "Vesuvius Eruption Baked Some People to Death—and Turned One Brain to Glass." *National Geographic* (January 23, 2020). URL: https://www.nationalgeographic.com/science/2020/01/vesuvius-baked-people-turned-brain-to-glass/.
7. Martyn, R. et al. "A Re-evaluation of Manner of Death at Roman Herculaneum Following the AD 79 Eruption of Vesuvius." *Antiquity* 94, 373 (2020). doi: 10.15184/aqy.2019.215.
8. Scarth, A. & Tanguy, J-C. *Volcanoes of Europe*. Oxford: Oxford University Press, 2001.
9. Halliday, Stephen. (2008) "London's Olympics, 1908." *History Today*. [Blog post] URL: https://www.historytoday.com/archive/london%E2%80%99s-olympics-1908.
10. Pratt, Sara E. "Benchmarks: March 17, 1944: The Most Recent Eruption of Mount Vesuvius." *Earth Magazine* (March 17, 2016). URL: https://www.earthmagazine.org/article/benchmarks-march-17-1944-most-recent-eruption-mount-vesuvius.
11. Billings, Malcolm. "Living Under a Volcano." *BBC News* (March 27, 2004). URL: http://news.bbc.co.uk/1/hi/programmes/from_our_own_correspondent/3502040.stm.
12. Satterthwaite, D.J. *Truth Flies with Fiction*. Bloomington, IN: Archway Publishing, 2014.
13. 57th Bomb Wing Association—War Diary of the 340th Bombardment Group March 1944. (2013) A Diary by Sergeant McRae. [Archival records] URL: http://57thbombwing.com/340th_History/340th_Diary/15_March1944.pdf.

ILLUSTRATION CREDITS

Prologue: The Gate in the Sky | *Artwork by Justin Estcourt*

I. The Fountain of Fire | *USGS Hawaiian Volcano Observatory (HVO)*

II. The Supervolcano | *Jim Peaco; June 22, 2006; Catalog #20386d; Original #IT8M4075 / National Park Service*

III. The Great Ink Well | *Kate Laxton*

IV. The Vaults of Glass | *Schmidt Ocean Institute and NOAA-PMEL Earth-Ocean Interactions Program*

V. The Pale Guardian | *NASA/JSC*

VI. The Toppled God | *NASA/JPL-Caltech*

VII. The Inferno | *NASA/JSC/JPL*

VIII. The Giant's Forge | *NASA/JPL/University of Arizona*

Epilogue: The Time Traveler | *Artwork by Justin Estcourt*

INDEX